MALLIAVIN CALCULUS FOR PROCESSES WITH JUMPS

STOCHASTICS MONOGRAPHS
Theory and Applications of Stochastic Processes

A series of books edited by Mark Davis, Imperial College, London, UK

Volume 1
Contiguity and the Statistical Invariance Principle
P. E. Greenwood and A. N. Shiryayev

Volume 2
Malliavin Calculus for Processes with Jumps
K. Bichteler, J. B. Gravereaux and J. Jacod

Additional volumes in preparation

ISSN: 0275-5785

This book is part of a series. The publishers will accept continuation orders which may be cancelled at any time and which provide for automatic billing and shipping of each title in the series upon publication. Please write for details.

MALLIAVIN CALCULUS FOR PROCESSES WITH JUMPS

KLAUS BICHTELER

The University of Texas at Austin

JEAN-BERNARD GRAVEREAUX

Université de Rennes

JEAN JACOD

Université Pierre et Marie Curie,
Paris

GORDON AND BREACH SCIENCE PUBLISHERS
New York London Paris Montreux Tokyo

Gordon and Breach Science Publishers

Post Office Box 786
Cooper Station
New York, New York 10276
United States of America

Post Office Box 197
London WC2E 9PX
England

58, rue Lhomond
75005 Paris
France

Post Office Box 161
1820 Montreux 2
Switzerland

14-9 Okubo 3-chome
Shinjuku-ku, Tokyo 160
Japan

Library of Congress Cataloging-in-Publication Data

Bichteler, Klaus.
Malliavin calculus for processes with jumps.

(Stochastic monographs ; v. 2)
Bibliography: p.
Includes index.
1. Stochastic analysis. 2. Functional analysis.
I. Gravereaux, Jean-Bernard, 1945– . II. Jacod, Jean. III. Title. IV. Series.
QA274.2.B53 1987 519.2 86-31825
ISBN 2-88124-185-9
ISSN 0275-5785

CONTENTS

Introduction to the series

The journal *Stochastics* publishes research papers dealing with stochastic processes and their applications in the modelling, analysis and optimization of systems subject to random disturbances. Stochastic models are now widely used in engineering, the physical and life sciences, economics, operations research, and elsewhere. Moreover, these models are becoming increasingly sophisticated and often stretch the boundaries of the theory as it exists. A primary aim of *Stochastics* is to further the development of the field by promoting an awareness of the latest theoretical developments on the one hand and of all problems arising in applications on the other.

In association with *Stochastics*, we are now publishing *Stochastics Monographs*, a series of independently produced volumes with the same aims and scope as the journal. *Stochastics Monographs* will provide timely and authoritative coverage of areas of current research in a more extended and expository form than is possible within the confines of a journal article. The series will include extended research reports, material derived from lecture courses on advanced topics, and multi-author works with a unified theme based on conference or workshop presentations.

MARK DAVIS

PREFACE

Since Malliavin introduced the new method in stochastic analysis, which now bears his name, much work has been done on various theoretical and applied aspects of the subject. However, essentially all this work has been concerned with analysis of continuous processes.

It was thus very tempting to see whether significant results could be achieved for discontinuous processes using the same sort of analysis, especially after Bismut cleared part of the way in 1983. We started by extending Bismut's approach in a relatively short paper, then discovered that the original approach of Malliavin, Stroock and others was also feasible for discontinuous processes. The two approaches were compared and this work grew into the present monograph. This book provides several new results, but the emphasis is clearly on methods.

JEAN JACOD

PREFACE

CHAPTER I

RESULTS

Section 1: INTRODUCTION

Malliavin [19] succeeded in proving some of Hörmander's regularity results using purely probabilistic techniques. Recall the

Problem 1: Under which conditions on the pair (a,B) of coefficients does the semi-group $(P_t(x,dy))_{t \geqq 0}$ with generator

$$L = \sum_i a^i(x)\frac{\partial}{\partial x_i} + \frac{1}{2}\sum_{i,j} B^{ij}(x)\frac{\partial^2}{\partial x_i \partial x_j} \tag{1-1}$$

have a density $p_t(x,y)$: $P_t(x,dy) = p_t(x,y)dy$? When is the density of class C^∞ in y or even in (x,y) or in (t,x,y) jointly?

The probabilistic argument is this. The measure $P_t(x,dy)$ is also the law of the solution X_t^x of the stochastic differential equation

$$X_t^x = x + \int_0^t a(X_s^x)ds + \int_0^t b(X_s^x)dW_s \tag{1-2}$$

where bb^T is the matrix B of 1-1 and W is an m-dimensional Wiener process (b^T denotes the transpose of b; we shall not worry here about the problems connected with the regularity of the representation $B=bb^T$). From the probabilistic angle the problem is therefore to discover under which conditions on the pair (a,b) of coefficients the distribution of the function X_t^x on the Wiener space has a density or even a smooth density.

Malliavin attacked this problem by transferring to

Wiener space the analysis one would use to solve the corresponding problem on $\mathbb{R}^d$. Integration-by-parts plays a central role in this approach (integration-by-parts formulae on Wiener spaces were actually known for some time: see Kuo [15] and Haussmann [11]). Malliavin calculus was developed and extended by several authors. Stroock [24], [25], [26] establishes the central integration-by-parts formula for the number operator, which is the generator of a diffusion semi-group on Wiener space (infinite-dimensional Ornstein-Uhlenbeck process). Similar approaches were given by Shigekawa [23] and Ikeda and Watanabe [12]. Stroock has also several regularity results not accessible to analytic treatment.

Bismut [6] uses Girsanov's theorem and flows - that is, deterministic semi-groups - on Wiener space to obtain a "directional" integration-by-parts formula. His method was simplified to some extent by Fonken [3], [4], [9] and Norris [21]. It has the advantage, recognized and exploited by Bismut himself [7], of generalizing rather easily to integro-differential operators of the form

$$L' = L + K \ ,$$
$$Kf(x) = \int [f(x+y) - f(x) - \sum_i \frac{\partial}{\partial x_i}(x) y_i] K(x,dy). \qquad (1\text{-}3)$$

Here K is a positive kernel on $\mathbb{R}^d$ that integrates the function $y \rightsquigarrow |y|^2$ (a "Lévy kernel"). Again a member $P'_t(x,dy)$ of the semi-group with generator L' is the law of the solution X^x_t of a stochastic differential equation, namely

$$X^x_t = x + \int_0^t a(X^x_s)ds + \int_0^t b(X^x_s)dW_s + \int_0^t \int_E c(X^x_{s-}, z)d\tilde{\mu} \qquad (1\text{-}4)$$

driven by time, Wiener process W and the "compensated Poisson measure" $\tilde{\mu}$ of a Poisson measure μ on $\mathbb{R}_+ \times E$:

E is an auxiliary space, with a positive σ-finite measure G, the intensity measure of μ is $\nu(dt,dx)=dt\times G(dx)$. Moreover, 1-3 and 1-4 are connected through

$$B = bb^T \; , \; K(x,A) = \int_E 1_{A\setminus\{0\}}(c(x,z))G(dz). \tag{1-5}$$

Bismut [7] can thus address:

<u>Problem 2</u>: Under which conditions on the triple (a,b,K) of 1-3 does $P'_t(x,dy)$ admit a density $p'_t(x,y)$, a regular density, a joint regular density?

Bismut solves this problem in a very special case where the Markov process with generator L' has a distribution which, for any starting point, is (by construction) absolutely continuous with respect to the distribution of a fixed process with stationary increments whose semi-group admits densities. The question of the existence of a density for (P'_t) does therefore not arise. The regularity of these densities, however, is a difficult problem. He [8] has also solved with his technique another, closely related, problem, namely when L' is the generator of a continuous diffusion with boundary; although the process is continuous, a Lévy kernel arises in connection with the excursion process (see also Léandre [18]).

Here we investigate a problem closely related to Problem 2, in a much more general context than Bismut; there is however a notable difference (and, as the non-probabilistically minded analyst might say, unfortunate):

<u>Problem 3</u>: Under which conditions on the triple (a,b,c) of 1-4 does $P'_t(x,dy)$ admit a density, a regular density, a joint regular density?

The main results are described in Section 2. We give a reasonably (?) general condition on (a,b,c) for the existence of a density: basically, it says that the diffuse part (expressed by b) and the jump part (expressed by c, or rather its derivatives $\partial c/\partial z_i$) must "fill" the whole tangent space at every point $x \in \mathbb{R}^d$; thus it essentially is a condition of *non-degeneracy*. When X^x is continuous (i.e. $c \equiv 0$) this amounts to non-degeneracy, or strict ellipticity, of the matrix $B = bb^T$: we are far from recovering the full force of Hörmander's Theorem (it would be possible however to get "weak Hörmander" conditions: see the works [17], [18] of Léandre, who obtains such results in a particular case). Then we state a *uniform non-degeneracy* condition which yields a given order of differentiability of the density.

In Section 3 we sketchily develop an example, closely related to [17].

Now, the emphasis of this monograph is not really put on the above-mentionned results, *but on the methods*: we apply and extend Bismut's method; we also extend Malliavin-Stroock's method, introducing to that effect a "Malliavin calculus" on Poisson space. It is worth mentioning that the two methods give essentially the same results (with a slight bonus for Bismut's one), as far as Problems 1 or 3 are concerned.

Chapter II is concerned with some useful techniques and more precisely two of these:

1) A general "integration-by-parts" setting is introduced in Section 4, and put to work for obtaining (smooth) densities of random variables. This is purely abstract, without reference to neither Poisson measures

nor Wiener processes (we advise to read §§3-a,b only).

2) Some rather general stability and differentiability properties for stochastic differential equations, in Section 5. The results are by no means novel, but they are adapted to our needs.

Chapter III is devoted to Bismut's method (we have already presented a version of the 1-dimensional case in [5]; see also [1], of course!): the "calculus of variations" is expounded in Section 6, while Section 7 contains the proofs of the main theorems.

Finally, Malliavin-Stroock's approach is presented in Chapter IV (some of the results have been announced in [10]). Malliavin's operators are presented in Section 8, in a rather abstract manner (not related to differential equations), and following Stroock [24], [25] rather closely. Sections 9 and 10 give applications to Wiener-Poisson space and to stochastic differential equations. Section 11 provides another proof of the main theorems, and Section 12 is devoted to comparing the two methods and to further comments.

Section 2: THE MAIN RESULTS

§2-a. GENERAL SETTING AND ASSUMPTIONS

The time interval is the bounded interval $[0,T]$. Despite first appearances, it will make our non-degeneracy condition easier to state and to interpret if we consider 1-4 with several Poisson measures driving, instead of only one - just as an m-dimensional Wiener process is but a collection of m independent 1-dimensional ones. Accordingly, we consider the

2-1 HYPOTHESES: $(\Omega, \underline{\underline{F}}, (\underline{\underline{F}}_t)_{t\in[0,T]}, P)$ *is a filtered space endowed with:*

- *a standard* m-*dimensional Wiener process* $W=(W^i)_{i\leq m}$;

- *for* $1\leq\alpha\leq A$, *a Poisson random measure* $\mu_\alpha = \mu_\alpha(\omega;dt,dz)$ *on* $[0,T]\times E_\alpha$, *where* E_α *is an open subset of* $\mathbb{R}^{\beta\alpha}$ *with infinite Lebesgue measure. The compensator* ν_α *of* μ_α *is of the form* $dt\times G_\alpha(dz)$, *where* G_α *denotes Lebesgue measure on* E_α. *The "compensated Poisson measure"* $\tilde{\mu}_\alpha$ *is given by* $\tilde{\mu}_\alpha=\mu_\alpha-\nu_\alpha$;

- *another Poisson measure* $\mu=\mu(\omega;dt,dx)$ *on* $[0,T]\times E$, *with intensity measure* $\nu(dt,dz)=dt\times G(dz)$; *here* G *is a positive* σ-*finite measure on a measurable space* $(E,\underline{\underline{E}})$ *and* $\tilde{\mu}=\mu-\nu$ *is the compensated measure;*

- *the random elements* (W^i,μ_α,μ) *are independent.*

Next, we are given a family of coefficients:

$$a = (a^i)_{1\leq i\leq d}: \mathbb{R}^d \to \mathbb{R}^d$$

$$b = (B^{ij})_{1\le i\le d,\,1\le j\le m} \;:\; \mathbb{R}^d \to \mathbb{R}^d\otimes\mathbb{R}^m$$
$$c_\alpha = (c_\alpha^i)_{1\le i\le d} \;:\; \mathbb{R}^d\times E \to \mathbb{R}^d$$
$$c = (c^i)_{1\le i\le d} \;:\; \mathbb{R}^d\times E \to \mathbb{R}^d.$$

For each $x\in\mathbb{R}^d$ we consider the following equation:

$$\begin{aligned} X_t^x = x &+ \int_0^t a(X_{s-}^x)ds + \int_0^t b(X_{s-}^x)dW_s \\ &+ \sum_\alpha \int_0^t\int_E c_\alpha(X_{s-}^x,z)\tilde{\mu}_\alpha(ds,dz) \\ &+ \int_0^t\int_E c(X_{s-}^x,z)\tilde{\mu}(ds,dz). \end{aligned} \tag{2-2}$$

The assumptions made below will ensure the existence and uniqueness of a solution X^x to 2-2 as an $\mathbb{R}^d$-valued càdlàg process (strong solution). In fact, X^x is a strong Markov process with generator:

$$L' = L + \sum_\alpha K_\alpha + K \qquad \text{with}$$

$$\begin{aligned} Lf(x) &= \sum a^i(x)\frac{\partial}{\partial x_i}f(x) + \frac{1}{2}\sum_{i,j} B^{ij}(x)\frac{\partial^2 f}{\partial x_i\partial x_j}(x) \\ K_\alpha f(x) &= \int\{f(x+y) - f(x) - \sum_i y_i\frac{\partial}{\partial x_i}f(x)\}K_\alpha(x,dy) \\ Kf(x) &= \int\{f(x+y) - f(x) - \sum_i y_i\frac{\partial}{\partial x_i}f(x)\}K(x,dy). \end{aligned} \tag{2-3}$$

The connection between the matrix B and kernels K_α, K on the one hand, and the coefficients of 2-2 on the other, is given by

$$\begin{aligned} B(x) &= b(x)b(x)^T \\ K_\alpha(x,A) &= \int_E 1_{A\setminus\{0\}}(c_\alpha(x,z))G_\alpha(dz) \\ K(x,A) &= \int_E 1_{A\setminus\{0\}}(c(x,z))G(dz). \end{aligned} \tag{2-4}$$

2-5 REMARK: It should be noted that 2-3 is not any

more general than 1-3; it is always possible to aggregate the Poisson measures μ_α and μ into a single Poisson measure on $\sum_\alpha E_\alpha + E$. By the same token, it is no restriction to assume that G_α is Lebesgue measure; any kernel K_α on $\mathbb{R}^d$ can be written in the form 2-4, provided G_α has infinite mass.

2-6 REMARK: In 2-2, dW_t means Ito-differential. This is contrary to the customary use of Stratonovitch differentials in the present context, but it is made necessary because the process X^x is not continuous.

In many formulae, vectors, matrices or higher order arrays appear; in general, we shall not show summation indices or symbols, unless there is ambiguity (for example in 2-2, the second integral really is $\sum_{1\le j\le m} \int_0^t b^{\cdot j}(X^x_{s-})dW^j_s$). If a is a vector or a matrix, a^T denotes its transpose. for any finite array a, $|a|^2$ is the sum of the squares of its entries; and $\det(a)$ denotes the determinant of the square matrix a.

We denote by $C^r_p(R^n)$ (resp. $C^r_b(R^n)$, resp. $C^r_o(R^n)$) the space of all r times continuously differentiable functions on $\mathbb{R}^n$, whose derivatives of all order between 0 and r are with polynomial growth (resp. bounded, resp. with compact support). If $f\in C^r_p(R^n)$, $D_x f$ stands for the *row* vector $(\partial f/\partial x_i)_{1\le i\le n}$, and further derivatives are denoted $D^m_{x^m} f$. If $f(\overline{x},\overline{z}) = f(x_1,..,x_d;z_1,..,z_n)$ is a differentiable function on $\mathbb{R}^d\times\mathbb{R}^n$ we denote by $D_x f$ and $D_z f$ the row vectors $(\partial f/\partial x_i)_{1\le i\le d}$ and $(\partial f/\partial z_i)_{1\le i\le n}$; further derivatives are $D^2_{x^2}f$, $D^2_{xz}f$,...,$D^{p+q}_{x^p z^q} f$.

In order to carry out the program outlined in the introduction, we shall need to differentiate a number of times. To do this, we clearly need some regularity on the coefficients of Equation 2-2. Here are the assumptions (below, $r \in \mathbb{N}^*$).

2-7 ASSUMPTION (A-r).

(i) a *and* b *are* r*-times differentiable with bounded derivatives of all order between* 1 *and* r.

(ii) c_α *is* r*-times differentiable on* $\mathbb{R}^d \times E$, *and*

$$c_\alpha(0,.) \in \cap_{2 \le p < \infty} L^p(E_\alpha, G_\alpha)$$

$$\sup_x |D^n_{x^n} c_\alpha(x,.)| \in \cap_{2 \le p < \infty} L^p(E_\alpha, G_\alpha) \quad \text{for } 1 \le n \le r$$

$$\sup_{x,z} |D^{n+q}_{x^n z^q} c_\alpha(x,z)| < \infty \quad \text{for } 1 \le n+q \le r \text{ and } q \ge 1.$$

(iii) c *is* $\underline{\underline{\mathbb{R}}}^d \times \underline{\underline{E}}$*-measurable;* $c(.,z)$ *is* r*-times differentiable, and*

$$c(0,.) \in \cap_{2 \le p < \infty} L^p(E,G)$$

$$\sup_x |D^n_{x^n} c(x,.)| \in \cap_{2 \le p < \infty} L^p(E,G) \quad \text{for } 1 \le n \le r.$$

We shall always assume at least (A-2), which will garantee that 2-2 has one and only one solution (unique up to evanescent sets): see e.g. Section 5 below, or [13].

We end this subsection fixing notation. $\underline{\underline{P}}$ denotes the predictable σ-field on $\Omega \times [0,T]$. We define the "maximal process" H^* of a process H by

$$H^*_t = \sup_{s \le t} |H_s|. \tag{2-8}$$

For stochastic integrals we use the standard notation of [13] or [20], for random measures those of [13]. The stochastic (indefinite) integral of a predictable row vector $H = (H^i)_{1 \le i \le m}$ with respect to W is written:

$$H * W = \sum_i H^i * W^i = \sum_i \int_0^{\cdot} H^i_s dW^i_s.$$

similarly, $H * t$ denotes the process

$$H*t(\omega) = \int_0^{\cdot} H_s(\omega)ds \ .$$

For a measurable function V on $\Omega\times[0,T]\times E$ we set

$$V*\mu_t(\omega) = \int_0^t \int_E V(\omega,s,z)\mu(\omega;ds,dz)$$

when this integral exists, and similarly for $V*\nu$, $V*\mu_\alpha$, $V*\nu_\alpha$. When V is $\underline{\underline{P}}\times\underline{\underline{E}}$-measurable, we can define the stochastic integral process $V*\tilde{\mu}$ of V relative to $\tilde{\mu}=\mu-\nu$ if and only if $(|V|^2\wedge|V|)*\nu_T<\infty$ a.s., and similarly for $V*\tilde{\mu}_\alpha$. With this notation, 2-2 reads

$$\begin{aligned} X^x = x + a(X^x_-)*t + b(X^x_-)*W \\ + \sum_\alpha c_\alpha(X^x_-)*\tilde{\mu}_\alpha + c(X^x_-)*\tilde{\mu} \end{aligned} \tag{2-9}$$

where $c_\alpha(X^x_-)$ stands for $c_\alpha(X^x_{t-}(\omega),z)$. We usually prefer 2-9 to 2-2 as it is somewhat more compact.

§2-b. EXISTENCE OF A DENSITY

In the sequel, a discriminating role is played by the following functions on $\mathbb{R}^d$ (resp. $\mathbb{R}^d\times E$). They take values in the set of symmetric nonnegative d×d matrices:

$$B(x) = b(x)b(x)^T, \tag{2-10}$$

$$C_\alpha(x,z) = \begin{cases} \{(I+D_xc_\alpha)^{-1}D_zc_\alpha D_zc_\alpha^T(I+D_xc_\alpha)^{-1,T}\}(x,z) \\ \qquad \text{if } I+D_xc_\alpha(x,z) \text{ is invertible,} \\ 0 \qquad \text{otherwise.} \end{cases}$$

Here, I is the d×d identity matrix. We may set

$$\begin{aligned} N_x &= \text{kernel of } B(x) \text{ in } \mathbb{R}^d \\ N^\alpha_{xz} &= \text{kernel of } C_\alpha(x,z) \text{ in } \mathbb{R}^d; \end{aligned} \tag{2-11}$$

equivalently,

N_x = orthocomplement of the image of $\mathbb{R}^m$ in $\mathbb{R}^d$ under the d×m matrix $b(x)$,

$$\left.\begin{array}{l} N^{\alpha}_{xz} = \text{orthocomplement of the image of } \mathbb{R}^{\beta_\alpha} \\ \quad \text{in } \mathbb{R}^d \text{ under the } d\times\beta_\alpha \text{ matrix} \\ \quad \{I+D_x c_\alpha(x,z)\}^{-1} D_z c_\alpha(x,z), \text{ and } N^{\alpha}_{xz}=\mathbb{R}^d \\ \quad \text{if } I+D_x c_\alpha(x,z) \text{ is not invertible.} \end{array}\right\} \quad (2-12)$$

The non-degeneracy assumption that will garantee the existence of a density is this:

2-13 <u>ASSUMPTION (B)</u>: *For each* $\alpha=1,\ldots,A$ *there is a Borel subset* Γ_α *of* $\mathbb{R}^d\times E_\alpha$ *such that, if* $\Gamma_{\alpha,x}$ *is the* x-*section of* Γ_α *in* E_α *and if*

$$W^{\alpha}_{x} = \begin{cases} \bigcup_{z\in\Gamma_{\alpha,x}} N^{\alpha}_{xz} & \text{if } G_\alpha(\Gamma_{\alpha,x})=+\infty \\ \mathbb{R}^d & \text{if } G_\alpha(\Gamma_{\alpha,x})<+\infty, \end{cases}$$

then $(\bigcap_\alpha W^{\alpha}_{x})\cap N_x = \{0\}$ *for all* $x\in\mathbb{R}^d$.

The first of our main theorems says:

2-14 <u>THEOREM</u>: *Under* (A-3) *and* (B), *the variables* X^x_t *have a density* $y \rightsquigarrow p_t(x,y)$ *for all* $x\in\mathbb{R}^d, t\in(0,T]$.

Note that the orthocomplement of N_x (resp. N^{α}_{xz}) is the vector subspace of $\mathbb{R}^d$ spanned by the eigenvectors of $B(x)$ (resp. $C_\alpha(x,z)$) corresponding to positive eigenvalues. Thus an equivalent, and perhaps more appealing way to state Assumption (B) is this:

2-15 <u>ASSUMPTION (B')</u>: $(\Leftrightarrow$(B)) *For each* $\alpha=1,\ldots,A$ *there is a Borel subset* Γ_α *of* $\mathbb{R}^d\times E_\alpha$ *with the following property: let* $I(x)=\{\alpha: G_\alpha(\Gamma_{\alpha,x})=+\infty\}$; *then for each* $x\in\mathbb{R}^d$ *and for each choice of* z_α *in* $\Gamma_{\alpha,x}$, *the vector space spanned by* $(N_x)^\perp$ *and the family* $\{(N^{\alpha}_{x,z_\alpha})^\perp\}_{\alpha\in I(x)}$ *equals* $\mathbb{R}^d$.

In other words, for each $x\in\mathbb{R}^d$ and each choice z_α

in $\Gamma_{\alpha,x}$ when $\alpha \in I(x)$, the symmetric nonnegative matrix $B(x) + \sum_{\alpha\in I(x)} C_\alpha(x,z_\alpha)$ is nondegenerate. The assumption $G_\alpha(\Gamma_{\alpha,x}) = +\infty$ implies that in every finite interval $(s,t]$ the Poisson measure μ_α has infinitely many jumps whose size belongs to $\Gamma_{\alpha,x}$. Phrased thus, it becomes clear that the true nature of Assumption (B) is a condition of *non-degeneracy* on the whole collection (B,C_α).

2-16 REMARK: This theorem, as well as the three other main theorems below, are proved in Section 7 via Bismut's approach. As a matter of fact, one can and will prove them under assumptions called (A'-r), which are slightly less stringent (and much more complicated to state) than (A-r): see 6-3.

Another proof of these theorems, via Malliavin-Stroock's approach, is provided in Section 11.

2-17 REMARK: Because of (A-r-(ii)), $\{I+D_x c_\alpha(x,z)\}^{-1}$ is as close as one wants to I, outside a set of G_α-finite measure. Hence $(N^\alpha_{xz})^\perp$ is "very close" to the image of $\mathbb{R}^{\beta_\alpha}$ under the linear map $D_z c_\alpha(x,z)$, this being the space in which the size $c_\alpha(x',z)$ of a jump of X^x at t is most likely to sit, given that $X^x_{t-}=x'$ and that μ_α has a point at (t,z). Thus it would have been somehow more intuitive if in condition (B), $(I+D_x c_\alpha)^{-1} D_z c_\alpha$ could be replaced by $D_z c_\alpha$ alone. Unfortunately, we could not achieve this substitution, due mainly to the fact that the set W^α_x in 2-13 is not closed, and in spite of the smallness of $D_x c_\alpha$.

2-18 REMARK: As seen above, (B) allows for a genuine interplay between W and the μ_α's. However, if $c_\alpha \equiv 0$ for all α then (B) reduces to nondegeneracy of $B(x)$ for all

x: this is rather crude, and far from "weak Hörmander conditions". A (relatively) simple addendum to the proof in Section 7 would indeed provide for conditions closer to Hörmander's conditions, as Léandre did in [17] in a particular case (namely, the case expounded in Section 3 below).

2-19 REMARK: (B) also shows the point in separating one driving Poisson measure μ from the others, the smoothness conditions on c being much weaker than on the c_α's. Actually it is often possible to chop an infinitesimal generator L' into pieces as in 2-3 with the c_α's that are very regular and a residual c that is less regular.

At this juncture, it should be noted that Bismut [7], in the particular case he studies, is asking for only two degrees of differentiability in z for the c_α's, whatever r is in (A-r). This is an important improvement upon (A-r), which comes from a difference in the "iteration procedure" which is basic to the method.

§2-c. REGULARITY OF THE DENSITY

We begin by introducing some functions that might look strange at first sight; we shall make a stab at explaining them in the next subsection.

2-20 DEFINITION: A measurable function $f_\alpha: E_\alpha \to [0,\infty)$ is called (ζ,θ)-*broad*, where ζ and θ are two positive numbers, if the following integral is finite:

$$\gamma_\alpha(\zeta,\theta,f_\alpha) = \int_0^\infty ds\ s^{\zeta-1}\ \exp\{-\theta\int_{E_\alpha}(1-e^{-sf_\alpha(z)})dz\}. \quad (2\text{-}21)$$

This means that f_α does not go too fast to 0 at

infinity. For example, we shall see soon that it implies

$$G_\alpha(f_\alpha>0) = +\infty. \tag{2-22}$$

For each α, we also consider a function ρ_α: $E_\alpha \to [0,\infty)$ with the following properties:

2-23 *Properties on* ρ_α: ρ_α is of class C_b^∞, and $\rho_\alpha(z) \to 0$ as z goes to the boundary of E_α, and $|D^r_{z^r}\rho_\alpha| \in L^1(E_\alpha, G_\alpha)$ for all $r \in \mathbb{N}$.

(Since E_α is a countable union of β_α-dimensional rectangles, it is always possible to find functions meeting 2-23 and being strictly positive).

2-24 ASSUMPTION (SB-(ζ,θ)): *There exist two constant* $\varepsilon>0$, $\delta\geq 0$, *and for all* $\alpha=1,\ldots,A$ *a* (ζ,θ)*-broad function* f_α *and a function* ρ_α *meeting 2-23, such that for all* $x,y \in \mathbb{R}^d$,

$$y^T B(x) y + \sum_\alpha \inf_{z: f_\alpha(z)>0} \frac{\rho_\alpha(z)}{f_\alpha(z)} y^T C_\alpha(x,z) y \geq |y|^2 \frac{\varepsilon}{1+|x|^\delta}. \tag{2-25}$$

This is evidently a *uniform non-degeneracy condition* (local uniform(!) would suit better). It clearly implies (B): simply take $\Gamma_\alpha = \mathbb{R}^d \times \{f_\alpha>0\}$ and use 2-22.

2-26 ASSUMPTION (SC): *There is a constant* $\zeta>0$ *such that* $|\det\{I+D_x c_\alpha(x,z)\}| \geq \zeta$ *and* $|\det\{I+D_x c(x,z)\}| \geq \zeta$ *identically.*

2-27 THEOREM: *Let* $r \in \mathbb{N}^*$ *and* $t \in (0,T]$. *Assume (SC) and either*

(i) (A-(r+d+3)) *and* (SB-(ζ,θ)) *with* $\theta \leq t$ *and*

$\zeta > \frac{2d(r+d+1)}{[t/\theta]}$ ($[\frac{t}{\theta}]$ denotes the integer part of $\frac{t}{\theta}$), *or*

(ii) (A-(r+3)) *and* (SB-(ζ,θ)) *with* $\theta \le t$ *and* $\zeta > \frac{2d^2(r+1)}{[t/\theta]}$.

Then, the density $y \rightsquigarrow p_t(x,y)$ *of the r.v.* X_t^x *exists and is of class* C^r.

2-28 THEOREM: *Let* $r \in \mathbb{N}^*$ *and* $t \in (0,T]$. *Assume (SC) and either*

(i) (A-(r+2d+3)) *and* (SB-(ζ,θ)) *with* $\theta \le t$ *and* $\zeta > \frac{2d(r+2d+1)}{[t/\theta]}$,

(ii) (A-(r+3)) *and* (SB-(ζ,θ)) *with* $\theta \le t$ *and* $\zeta > \frac{4d^2(r+1)}{[t/\theta]}$.

Then $(x,y) \rightsquigarrow p_t(x,y)$ *is of class* C^r.

2-29 THEOREM: *Let* $r \in \mathbb{N}^*$ *and* $t \in (0,T]$.

(i) Assume (A-(2r+4d+6)) *and* (SB-(ζ,θ)) *with* $\theta \le t$ *and* $\zeta > \frac{4d(r+2d+2)}{[t/\theta]}$; *assume also the following* (stronger than (SC)):

$$|\det\{I+uD_x c_\alpha(x,z)\}| \ge \zeta', \quad |\det\{I+uD_x c(x,z)\}| \ge \zeta' \qquad (2\text{-}30)$$

for all $u \in [0,1]$, *where* ζ' *is a constant. Then* $(s,x,y) \rightsquigarrow p_s(x,y)$ *is of class* C^r *on* $[t,T] \times \mathbb{R}^d \times \mathbb{R}^d$.

(ii) Assume (A-(2r+4)), *and* $c_\alpha \equiv 0$, $c \equiv 0$, *and* $y^T B(x) y \ge \varepsilon |y|^2/(1+|x|^\delta)$ *identically for some constants* $\varepsilon > 0$, $\delta \ge 0$. *Then* $(s,x,y) \rightsquigarrow p_s(x,y)$ *is of class* C^r *on* $[t,T] \times \mathbb{R}^d \times \mathbb{R}^d$.

A (ζ,θ)-broad function is (ζ',θ')-broad for $\zeta' \le \zeta$ and $\theta' \ge \theta$: this is consistent with the form of conditions (i) and (ii) in 2-27, 2-28 and 2-29. The broadness condition in 2-27-(i), say, becomes more stringent as

r increases and as t decreases. In fact, one could read the theorem as follows: suppose that (SB-(ζ,θ)) holds for some ζ,θ; under suitable regularity conditions the density of X_t^x, with $t\geq\theta$, is r times differentiable for all integers r with $r<[\frac{t}{\theta}]\zeta/2d-d-1$. Or also: given $r\in\mathbb{N}^*$, the density of X_t^x is r times differentiable whenever t is large enough to satisfy $[t/\theta]>2d(r+d+1)/\zeta$ and $t\geq\theta$.

The phenomenon of regularity increasing with t has already been encountered by Bismut [7]. As a matter of fact, the same phenomenon also occurs for the existence of a density (although our method is too crude to show this): one can construct a process X with stationary independent increments, starting at 0, such that X_1 has a singular distribution and X_2 has a density ([22],[28]).

Note that the assumptions (i) and (ii) in these theorems are not, in general, comparable, because the broadness condition in (ii) is stronger than the one in (i) (except for $d=1$, or $d=2$ and $r=0$).

§2-d. BROAD FUNCTIONS

We devote this subsection to some comments and some examples of broad functions. Firstly, the following statements are obvious:

2-31 Any function bigger than a (ζ,θ)-broad function is (ζ,θ)-broad itself (if it is measurable, of course).

2-32 If f_α is (ζ,θ)-broad, then so is γf_α for $\gamma>0$.

2-33 Any strictly positive constant function on E_α is

(ζ,θ)-broad, for all $\zeta>0,\theta>0$ (recall that $G_\alpha(E_\alpha)=+\infty$ by hypothesis.

2-34 LEMMA: *If* $G_\alpha(C)<\infty$, *then* f_α *is* (ζ,θ)-*broad if and only if* $f_\alpha 1_{C^c}$ *is* (ζ,θ)-*broad* (this obviously implies 2-22).

Proof. We have $-G_\alpha(C)\leq -\int_C (1-e^{-sf_\alpha(z)})dz$, hence

$$e^{-\theta G_\alpha(C)} \gamma_\alpha(\zeta,\theta,f_\alpha 1_{C^c})$$
$$= \int_0^\infty s^{\zeta-1}\, ds \exp\{-\theta G_\alpha(C) - \theta\int_{C^c} (1-e^{-sf_\alpha(z)})dz$$
$$\leq \gamma_\alpha(\zeta,\theta,f_\alpha),$$

while $\gamma_\alpha(\zeta,\theta,f_\alpha)\leq\gamma_\alpha(\zeta,\theta,f_\alpha 1_{C^c})$ is obvious.

2-35 *Example of broad functions on* $\mathbb{R}$: Let $\beta_\alpha=1$ and $E_\alpha=(\eta,\infty)$ or $E_\alpha=(-\infty,\eta)$ for some $\eta\in\mathbb{R}$. Then one easily checks that

$$f_\alpha(z) = |z|^\gamma e^{-\delta|z|} \quad \text{is } (\zeta,\theta)\text{-broad iff } \delta<\frac{\theta}{\zeta}, \gamma\in\mathbb{R}.$$

In particular, $|z|^\gamma$ is (ζ,θ)-broad for all $\zeta>0$, $\theta>0$, for any real γ. When $E_\alpha=\mathbb{R}$, we have the same result, with $\delta<\frac{2\theta}{\zeta}$ instead of $\delta<\frac{\theta}{\zeta}$.

2-36 *Example of broad functions in higher dimension:* Let $\beta_\alpha\geq 2$, and suppose that the boundary of E_α consists in a finite number of hyperplanes. Then $f_\alpha(z)=|z|^\gamma e^{-\delta|z|}$ is (ζ,θ)-broad in exactly three cases:

(i) $\delta<\frac{\theta}{\zeta}$ and E_α is contained in a cylinder with $(\beta_\alpha-1)$-dimensional bounded basis, and extends to infinity on one side of the basis;

(ii) $\delta<\frac{2\theta}{\zeta}$ and E_α is as above, but extends to infinity on both sides of the basis;

(iii) $\delta \in \mathbb{R}$, $\gamma \in \mathbb{R}$, and E_α is contained in no cylinder with bounded $(\beta_\alpha - 1)$-dimensional basis.

(Observe that 2-35 is but a special case of (i) or (ii)).

2-37 <u>COROLLARY</u>: *Suppose that each* E_α *is limited by a finite number of hyperplanes in* $\mathbb{R}^{\beta_\alpha}$. *In order that* (SB-$(\zeta,\theta)$) *be met, it suffices that there exist* $\varepsilon > 0$, $\delta \geq 0$, *and for each* $\alpha = 1,\dots,A$ *a real number* δ_α *such that* $\delta_\alpha < \theta/\zeta$ *(resp.* $\delta_\alpha < 2\theta/\zeta$, *resp.* $\delta_\alpha \in \mathbb{R}$*) if* E_α *is like in 2-36-(i) (resp. (ii), resp. (iii)), with*

$$y^T B(x) y + \sum_\alpha \inf_{z \in E_\alpha} (y^T C_\alpha(x,z) y) e^{\delta_\alpha |z|} \geq |y|^2 \frac{\varepsilon}{1 + |x|^\delta} . \qquad (2\text{-}38)$$

<u>Proof</u>. Let $E'_\alpha = E_\alpha \cap \{z \in \mathbb{R}^{\beta_\alpha} : |z| \geq 1\}$. One can easily find a function $\rho_\alpha : E_\alpha \to [0,\infty)$ meeting 2-23, such that $\rho_\alpha(z) = |z|^{-\beta_\alpha - 1}$ if $z \in E'_\alpha$. Then, due to 2-36 and 2-34, $f_\alpha(z) = e^{-\delta_\alpha |z|} \rho_\alpha(z)$ is (ζ,θ)-broad, and 2-25 follows from 2-38.

Section 3: ONE EXAMPLE

We give here an example of how the main theorems can be put to work on the original Problem 2, when the operator L' is given by 2-3, with the kernels K_α sitting on smooth curves. This example is essentially the case examined by Léandre in [18], but the results presented here are slightly different, and essentially weaker.

More precisely, $K_\alpha(x,.)$ admits a smooth curve $\Gamma_\alpha(x)$ in $\mathbb{R}^d$ as its support, on which it is absolutely continuous with respect to the arc-length measure. We suppose that the curves $\Gamma_\alpha(x)$ are nicely parametrized by $[0,1]$:

3-1 There is a C^1 mapping $y_\alpha:\mathbb{R}^d\times[0,1] \to \mathbb{R}^d$ such that $\Gamma_\alpha(x)$ is the image of $[0,1]$ under $t \rightsquigarrow y_\alpha(x,t)$. We suppose that $y_\alpha(x,0)=0$ and that $y_\alpha(x,t)\neq 0$ for $t>\gamma_\alpha(x):=\inf(s:\ y_\alpha(x,s)\neq 0)$, and that $D_t y_\alpha(x,\gamma_\alpha(x))\neq 0$.

Hence the curve leaves the origin at time $t=\gamma_\alpha(x)$ and never returns (as the reader will immediately notice, these are no restrictions for the problem at hand); also that $D_t y_\alpha(x,\gamma_\alpha(x))\neq 0$ can always be achieved by a suitable parametrization. Assume also that

$$K_\alpha(x,C) = \int_0^1 1_{C\setminus\{0\}}(y_\alpha(x,t))g_\alpha(x,t)dt \qquad (3-2)$$

where:

3-3 g_α is of class C^1 on $\mathbb{R}^d\times(0,1]$ and $\int_0^1 g_\alpha(x,t)dt=\infty$ for all x and $g_\alpha(x,t)\geq\zeta$ for all x,t and some constant $\zeta>0$.

In 3-3 the assumption $g_\alpha\geq\zeta$ is a most unfortunate restriction (see [5] for a discussion of this point in a similar context). The assumption that $\int_0^1 g_\alpha(x,t)dt=\infty$ is <u>not</u> a restriction, and does not mean that $K_\alpha(x,\mathbb{R}^d)=\infty$ but merely implies the equivalence

$$K_\alpha(x,\mathbb{R}^d) = \infty \quad \Leftrightarrow \quad \gamma_\alpha(x) = 0. \tag{3-4}$$

Now to apply our theorems we must obtain functions c_α satisfying 2-4 with suitable E_α. A simple way goes as follows. Set

$$\hat{g}_\alpha(x,z) = \int_z^1 g_\alpha(x,s)ds, \qquad z\in[0,1]. \tag{3-5}$$

This is a C^1 function on $\mathbb{R}^d\times(0,1]$ and, because of 3-3, $\hat{g}_\alpha(x,.)$ is a bijection of $(0,1]$ onto $[0,\infty)$. We denote by $\hat{g}'_\alpha(x,.)$ its inverse: $\hat{g}'_\alpha(x,\hat{g}_\alpha(x,z))=z$ for $z\in(0,1]$. Then set

$$E_\alpha = (0,\infty), \qquad c_\alpha(x,z) = y_\alpha(x,\hat{g}'_\alpha(x,z)). \tag{3-6}$$

A simple computation shows that 2-4 holds, and that

$$\begin{aligned} D_z c_\alpha(x,z) &= -\left(\frac{D_t y_\alpha}{g_\alpha}\right)(x,\hat{g}'_\alpha(x,z)) \\ D_x c_\alpha(x,z) &= D_x y_\alpha(x,\hat{g}'_\alpha(x,z)) \\ &\quad + \int_{\hat{g}'_\alpha(x,z)}^1 D_x g_\alpha(x,s)ds \left(\frac{D_t y_\alpha}{g_\alpha}\right)(x,\hat{g}'_\alpha(x,z)). \end{aligned} \tag{3-7}$$

Finally we are also given a,b,c as in 2-4. We will use the assumptions (A-r) and (SC) of §§2-b,c, the corresponding conditions on (y_α,g_α) being easily (although tediously!) obtained via 3-7. Observe that the last condition in 3-3 plays a central role in obtaining

(A-r)-(ii) for c_α (see [5] for a more detailed discussion of these assumptions.)

3-8 THEOREM: *Assume* (A-3) *with the* K_α*'s as above. Assume also that for each* $x\in\mathbb{R}^d$ *the vector space spanned by all eigenvectors of* $B(x) = b(x)b(x)^T$ *corresponding to positive eigenvalues, together with all the tangent vectors to the curves* $\Gamma_\alpha(x)$ *at the origin for those* α *for which* $K_\alpha(x,\mathbb{R}^d)=+\infty$, *is equal to* $\mathbb{R}^d$. *Then* X_t^x *admits a density* $p_t(x,.)$ *for all* $t>0$, $x\in\mathbb{R}^d$.

Proof. By 2-14 it suffices to prove that (B') (see 2-15) holds. With the notation 2-12, N^α_{xz} is the orthocomplement of the vector $U_\alpha(x,z)=$ $(I+D_xc_\alpha(x,z))^{-1}\, D_zc_\alpha(x,z)$ if $I+D_xc_\alpha$ is invertible, and $U_\alpha=0$ otherwise.

Let x be such that $K_\alpha(x,\mathbb{R}^d)=\infty$. The function $k(z)= D_xc_\alpha(x,z)$ is in $L^2((0,\infty),dz)$ and it has bounded derivatives, hence $\lim_{z\uparrow\infty}k(z)=0$. Moreover 3-7 yields $(D_zc_\alpha/|D_zc_\alpha|)(x,z) = -(D_ty_\alpha/|D_ty_\alpha|)(x,\hat{g}'_\alpha(x,z))$ if $D_ty_\alpha(x,\hat{g}'_\alpha(x,z)) \neq 0$; we also have $\lim_{z\uparrow\infty}\hat{g}'_\alpha(x,z)=0$ and $\gamma_\alpha(x)=0$ (see 3-4) and $D_ty_\alpha(x,0)\neq 0$. Using the definition of U_α and the property $k(z)\to 0$ as $z\uparrow\infty$, we obtain that $U_\alpha(x,z)\neq 0$ for z big enough and that the unit vector $U_\alpha/|U_\alpha|$ has:

$$\lim_{z\uparrow\infty} \frac{U_\alpha}{|U_\alpha|}(x,z) = \text{outward unit tangent vector to } \Gamma_\alpha(x) \text{ at } 0 . \tag{3-9}$$

Now, our assumptions and 3-9 imply that there are measurable functions $\varepsilon_\alpha:\mathbb{R}^d\to(0,\infty)$ such that for any choice of $z_\alpha\in(\varepsilon_\alpha(x),\infty)$ the vector space spanned by $(N_x)^\perp$ and the $U_\alpha(x,z_\alpha)$, with $\alpha\in I(x):=\{\alpha:K_\alpha(x,\mathbb{R}^d)=\infty\}$ is equal to $\mathbb{R}^d$. Therefore (B') is met with Γ_α =

$\{(x,z)\in\mathbb{R}^d\times E_\alpha : \alpha\in I(x) \text{ and } z>\varepsilon_\alpha(x)\}$.

Next, we give a smoothness result, aiming to a sufficient condition that pertains essentially to the geometrical properties of the curves $\Gamma_\alpha(x)$. The following *uniform non-degeneracy condition* is of this description:

3-10 <u>ASSUMPTION (SB')</u>: *There is* $\varepsilon>0$ *such that for all* $x,y\in\mathbb{R}^d$,

$$y^T B(x)y + \sum_{\alpha:\gamma_\alpha(x)=0} |y^T D_t y_\alpha(x,0)|^2 \geq \varepsilon|y|^2.$$

3-11 <u>ASSUMPTION (D)</u>: *For each* $\eta>0$ *there is a constant* $\rho(\eta)$ *such that:* $|D_t y_\alpha(x,\hat{g}'_\alpha(x,z)) - D_t y_\alpha(x,0)|\leq\eta$ *whenever* $\gamma_\alpha(x)=0$ *and* $z\geq\rho(\eta)$ (note that $D_t y_\alpha(x,\hat{g}'_\alpha(x,z)) \to D_t y_\alpha(x,0)$ as $z\uparrow\infty$ when $\gamma_\alpha(x)=0$; so this is an assumption on the uniformity in x for this convergence).

3-12 <u>THEOREM</u>: *Assume (A-(r+3)), (SB'), (D), (SC). Assume also that*

$$z\geq\zeta' \Rightarrow g_\alpha(x,\hat{g}'_\alpha(x,z)) \leq \zeta'' e^{\zeta z} \tag{3-13}$$

for three constants ζ, ζ', $\zeta''>0$. *Then* X_t^x *admits a density* $y \rightsquigarrow p_t(x,y)$ *of class* C^r, *provided* $t>4d^2(r+1)$; *moreover* p_t *is of class* C^r *in* (x,y), *provided* $t>8d^2(r+1)$.

<u>Proof</u>. We will prove that 2-38 holds with $\delta_\alpha=2\zeta$: the result will then readily follow from 2-37, and 2-27-(ii) or 2-28-(ii) (take $\theta=t$ in those). In fact, we will prove 2-38, with the infimum being taken on $E'_\alpha=(\phi,\infty)$ for some $\phi\in\mathbb{R}_+$ to be chosen, instead of E_α: because of 2-34 this is of no consequence.

For simplicity, set $V_\alpha(x):=D_t y_\alpha(x,0)$, and $\psi:=$

$\sup_{x,\alpha}|V_\alpha(x)|$, which is finite. We have seen in 3-8 that $\lim_{z\uparrow\infty} D_z c_\alpha(x,z)=0$, and the same argument (using (A-2)) indeed yields that $\lim_{z\uparrow\infty} \sup_x |D_z c_\alpha(x,z)|=0$. Hence if $\Delta_\alpha=(I+D_x c_\alpha)^{-1} - I$, there is a $\phi_1>0$ such that

$$z\geq\phi_1 \Rightarrow \|\Delta_\alpha(x,z)\| \leq \frac{1}{4\psi}\sqrt{\frac{\varepsilon}{2A}} \wedge \frac{1}{4} \tag{3-14}$$

where $\|.\|$ denotes the operator norm. Let U_α be as in the proof of 3-8. Then 3-7 yields

$$y^T U_\alpha(x,z) = g_\alpha(x,\hat{g}'_\alpha(x,z))^{-1}\{ y^T V_\alpha(x)+y^T\Delta_\alpha(x,z)V_\alpha(x)$$
$$+ y^T(I+\Delta_\alpha(x,z))[D_t y_\alpha(x,\hat{g}'_\alpha(x,z))-V_\alpha(x)]\}.$$

Thus 3-14 and (D) yield for $\phi_2 = \phi_1 \wedge \rho(\frac{1}{5}\sqrt{\frac{\varepsilon}{2A}})$:

$$\gamma_\alpha(x)=0, z\geq\phi_2 \Rightarrow \tag{3-15}$$
$$|y^T U_\alpha(x,z)| \geq g_\alpha(x,\hat{g}'_\alpha(x,z))^{-1}\{|y^T V_\alpha(x)|-\tfrac{1}{2}|y|\sqrt{\tfrac{\varepsilon}{2A}}\}.$$

Now we fix $x,y\in\mathbb{R}^d$. If $y^T B(x)y<\varepsilon|y|^2/2$ there is at least one α with $\gamma_\alpha(x)=0$ and $|y^T V_\alpha(x)|^2\geq\varepsilon|y|^2/2A$, by (SB'). Thus 3-15 results in

$$z\geq\phi_2 \Rightarrow |y^T U_\alpha(x,z)|\geq g_\alpha(x,\hat{g}'_\alpha(x,z))^{-1}\frac{|y|}{2}\sqrt{\frac{\varepsilon}{2A}}$$

and 3-13 yields, if $\phi=\zeta'\wedge\phi_2$:

$$z\geq\phi \Rightarrow |y^T U_\alpha(x,z)|^2 \geq \frac{\varepsilon}{8A\zeta''^2} e^{-2\zeta z} |y|^2$$

for some α, if $y^T B(x)y<\varepsilon|y|^2/2$. Therefore, for all $x,y\in\mathbb{R}^d$,

$$y^T B(x)y + \sum_\alpha \inf_{z\geq\phi} y^T C_\alpha(x,z)y\, e^{2\zeta z}$$
$$\geq \frac{\varepsilon}{2}\left(1 \wedge \frac{1}{4A\zeta''^2}\right).|y|^2.$$

So we have 2-38 and the proof is finished.

3-16 <u>REMARK</u>: 3-13 looks strange, as one would rather think that the bigger the density is, the more regularity one gets (because a "big" density means that there are many jumps). But in fact, 3-13 does not say "g_α is not too big", but rather "g_α is regular enough".

CHAPTER II

TECHNIQUES

We are going in this chapter to introduce two different tools which we need later.

The first one consists in a systematic exploitation of ideas of Malliavin [19] to the end of proving existence and smoothness of the density for the law of a mutidimensional random variable. It takes the form of an abstract formalism which will perhaps appear as mere nonsense: its sole interest lies in the fact that it avoids repeating twice the same argument, for Bismut's and for Malliavin-Stroock's approach.

The second tool consists in various stability and differentiability results for stochastic differential equations depending on a parameter. These results are more or less well known, although the presence of a Poisson measure complicates the matter (in fact, essentially the statements, because the proofs are the same as for the Wiener component). Naturally, we do not provide a complete theory, but we give only the strictly needed results: so the interest of Section 5 indeed lies in the fact that the results are given under the exactly suitable form.

Section 4: TOWARD EXISTENCE AND SMOOTHNESS OF THE DENSITY FOR A RANDOM VARIABLE

Consider a d-dimensional random variable $\Phi=(\Phi^i)_{i\leq d}$ on $(\Omega,\underline{\underline{F}},P)$. Our basic criterion for the law of Φ to admit a density is due to Malliavin [19] and is a very simple application of Fourier analysis; it constitutes the first part of the following lemma (see also Ikeda and Watanabe [12]), while the second part is a refinement due to Kosuoka and Stroock [16]. This lemma can also be considered as a variant of Sobolev's lemma.

4-1 LEMMA: *a) If there are two constants* $C>0$ *and* $n\in\mathbb{N}^*$ *such that for all* $f\in C_o^\infty(R^d)$ *and every multi-index* α *with* $|\alpha|\leq n$,

$$|E[D^\alpha_{x^\alpha} f(\Phi)]| \leq C\|f\|_\infty \tag{4-2}$$

then: (i) if $n=1$, Φ *admits a density;*

(ii) if $n\geq d+1$, Φ *admits a density of class* C^{n-d-1}.

b) Let $n\in\mathbb{N}^*$. *If for every multi-index* α *with* $|\alpha|\leq n$ *there is a random variable* $\Psi_\alpha\in L^{d+\varepsilon}$, *where* $\varepsilon>0$, *such that for all* $f\in C_o^\infty(R^d)$

$$E[D^\alpha_{x^\alpha} f(\Phi)] = E[f(\Phi)\Psi_\alpha], \tag{4-3}$$

then Φ *admits a density of class* C^{n-1}.

The only sensible way to get 4-2 seems to be proving 4-3 for some $\Psi_\alpha\in L^1$: so, going from (a) to (b), we gain on the order of differentiability (from $n-d-1$ to n-1), and we lose on the integrability assumption on Ψ_α (from L^1 to $L^{d+\varepsilon}$).

The question now is: how to get 4-3? the idea is of course to derive 4-3 for $|\alpha|=1$, and then to iterate the formula so obtained. Obviously 4-3 for $|\alpha|=1$ is an "integration-by-parts" formula.

§4-a. INTEGRATION-BY-PARTS SETTING

In this subsection, we fix the r.v. $\Phi=(\Phi^1,..,\Phi^d)$ of interest.

4-4 DEFINITION: An *integration-by-parts setting for* Φ is: (i) a r.v. $\sigma_\Phi=(\sigma_\Phi^{ij})_{1\le i,j\le d}$ with values in $\mathbb{R}^d\otimes\mathbb{R}^d$;

(ii) a r.v. $\gamma_\Phi=(\gamma_\Phi^i)_{1\le i\le d}$ with values in $\mathbb{R}^d$, and with $\gamma_\Phi^i\in\cap_{p<\infty}L^p$;

(iii) a linear space H_Φ of random variables, contained in $\cap_{p<\infty}L^p$, and which is stable under C_p^2: that is, if $\Psi=(\Psi^1,..,\Psi^n)\in(H_\Phi)^n$ and $F\in C_p^2(R^n)$, then $F(\Psi)\in H_\Phi$;

(iv) d linear maps $\delta_\Phi^j: H_\Phi \to \cap_{p<\infty}L^p$ such that if $\Psi=(\Psi^1,..,\Psi^n)\in(H_\Phi)^n$ and $F\in C_p^2(R^n)$, then

$$\delta_\Phi^j(F\circ\Psi) = \sum_i \frac{\partial F}{\partial x_i}(\Psi)\,\delta_\Phi^j(\Psi^i); \tag{4-5}$$

(v) for every $f\in C_p^2(R^d)$ and every $\Psi\in H_\Phi$, we have

$$E[\Psi \sum_i \frac{\partial F}{\partial x_i}(\Phi)\sigma_\Phi^{ij}] = E[f(\Phi)\{\Psi\gamma_\Phi^j+\delta_\Phi^j(\Psi)\}]. \tag{4-6}$$

At this point, it would seem simpler to merely introduce the sum $G_\Phi(\Psi)^i = \Psi\gamma_\Phi^i+\delta_\Phi^i(\Psi)$: the reason for singling out two distinct terms will become evident when we start iterating 4-6.

For the moment, we observe that 4-6 naturally leads to 4-3 for $|\alpha|=1$. Indeed, suppose that σ_Φ is invertible and that the entries of its inverse σ_Φ^{-1} all belong to $\mathcal{H}_\Phi$. Applying 4-6 to $\Psi^{ij}=(\sigma_\Phi^{-1})^{ij}$ and to the ith component $[D_x f(\Phi)\sigma_\Phi]^i = \sum_{k\leq d} \frac{\partial}{\partial x_k} f(\Phi)\sigma_\Phi^{ki}$ and summing up on i, yield

$$E(\frac{\partial}{\partial x_j} f(\Phi)) = E[f(\Phi) \sum_i G_\Phi((\sigma_\Phi^{-1})^{ij})^i],$$

which is 4-3 for $|\alpha|=1$.

A less crude way of getting rid of the "nuisance" term σ_Φ in 4-6 is proposed in the next theorem (essentially due to Malliavin [19] and Shigekawa [23]):

4-7 <u>THEOREM</u>: *Assume 4-4 and that* $\sigma_\Phi^{ij}\in\mathcal{H}_\Phi$ *for all* $i,j\leq d$. *If* P_1 *denotes the restriction of the measure* P *to the set where the matrix* σ_Φ *is invertible, the law of* Φ *under* P_1 (i.e., the image $P_1\circ\Phi^{-1}$ of P_1 on $\mathbb{R}^d$) *admits a density on* $\mathbb{R}^d$.

In particular, if σ_Φ *is* P-*a.s. invertible, the law of* Φ *under* P *admits a density on* $\mathbb{R}^d$.

<u>Proof</u>. Consider the following function $\phi:\mathbb{R}^d\otimes\mathbb{R}^d\to\mathbb{R}^d\otimes\mathbb{R}^d$, which clearly is of class C_b^∞:

$$\phi(x) = \begin{cases} x^{-1} \exp\text{-}\det(x)^{-2} & \text{if } \det(x)\neq 0 \\ 0 & \text{if } \det(x)=0. \end{cases} \tag{4-8}$$

By 4-4-(iii), $\phi^{ij}(\sigma_\Phi)$ belongs to $\mathcal{H}_\Phi$. Applying 4-6 to $\Psi^{ij}=\phi^{ij}(\sigma_\Phi)$ and $[D_x f(\Phi)\sigma_\Phi]^i$ and summing up on i, we get

$$E[\frac{\partial}{\partial x_j} f(\Phi) \exp\text{-}\det(\sigma_\Phi)^{-2}] = E(f(\Phi)Z_j) \tag{4-9}$$

where $Z_j=\sum_{i\leq d} G_\Phi(\phi^{ij}(\sigma_\Phi))^i$ belongs to L^1. If P' is the finite measure defined by $P'(d\omega) = P(d\omega).\exp\text{-}\det(\sigma_\Phi(\omega))^{-2}$, 4-9 yields $|E'(\frac{\partial}{\partial x_j} f(\Phi))|\leq C\|f\|_\infty$ with $C:=\sup_{j\leq d} E(|Z_j|)$. Then Lemma 4-1-a implies that the law of Φ under P' admits a density; furthermore $P_1 \sim P'$ by construction, so the first claim follows. The last claim is then obvious.

Before deriving other consequences, let us emphazise that a good part of what follows is concerned with obtaining an integration-by-parts setting for the variables of interest, namely $\Phi=X_t^x$ for $t\in(0,T]$ where X^x is the solution to 1-4. This is where the two methods diverge:

1) In Bismut's method, the construction 4-4 strongly depends upon the solution itself;

2) In Malliavin-Stroock's method, the construction 4-4 is more intrinsic, that is H_Φ does not depend on Φ (for "reasonable" Φ's, including the X_t^x above-mentioned).

§4-b. ITERATION OF THE INTEGRATION-BY-PARTS FORMULA

Again Φ is fixed here, and to simplify the notation we drop the subsript "Φ" in our given integration-by-parts setting for Φ, thus writing (σ,γ,H,δ).

Iterating 4-6 means applying 4-6 again to $\Psi'=G(\Psi) := \Psi\gamma+\delta(\Psi)$, which should then be in H for all Ψ that are needed. This motivates the present inductive construction of families of random variables:

$$\left.\begin{array}{l} C_0 \text{ is a } \textit{finite} \text{ set of } \mathbb{R}\text{-valued r.v., including } \gamma^i \text{ and } \sigma^{ij} \text{ for } i,j\leq d \\ C_{r+1} = C_r \cup \{\delta^i(\Psi):\ i\leq d, \Psi\in C_r\} \quad \text{if } C_r\subset H. \end{array}\right\} \quad (4\text{-}10)$$

For each r such that C_r is well-defined, we write Y_r for the multidimensional r.v. whose components are successively all the $\mathbb{R}$-valued r.v. that constitute C_r, and we call E_r the space in which Y_r takes its values: thus if C_0 contains $\alpha(0)$ distinct real r.v. ($\alpha(0)\geq d+d^2$) then

$$E_0 = \mathbb{R}^{\alpha(0)}, \quad E_{r+1} = E_r \times (\mathbb{R}^d \otimes E_r) \tag{4-11}$$

and the dimension $\alpha(r)$ of E_r is $\alpha(r)=\alpha(0)(1+d)^r$.

At first, we derive an integration-by-parts formula which is more tractable than 4-6.

The function $g(x)=\det(x):\mathbb{R}^d\otimes\mathbb{R}^d \to \mathbb{R}$ is a polynomial; there is another polynomial $h:\mathbb{R}^d\otimes\mathbb{R}^d \to \mathbb{R}^d\otimes\mathbb{R}^d$ such that $x^{-1}=h(x)/g(x)$ whenever x is invertible.

Assume that $C_\ell \subset H$ for some $\ell\in\mathbb{N}$, and let $q\geq 2$. Let $F\in C_p^q(E_\ell)$ (i.e. a function of class C^q on E_ℓ, whose derivatives of all order have at most polynomial growth). We observe that $g(\sigma)$, $h(\sigma)$, γ, $\delta^j(g\circ\sigma)$, $\delta^j(h^{ji}\circ\sigma)$ are polynomial functions of the components of $Y_{\ell+1}$ (use 4-5 for the last two functions); using 4-5 once more, we see that $\delta^j(F\circ Y_\ell) = G^j(Y_{\ell+1})$, where $G^j\in C_p^{q-1}(E_{\ell+1})$. Hence the formula

$$\begin{aligned}\Delta_F^i(Y_{\ell+1}) = \textstyle\sum_{j\leq d} \{F(Y_\ell)[\, g(\sigma)h^{ji}(\sigma)\gamma^j \\ + \delta^j(g\circ\sigma)h^{ji}(\sigma) + g(\sigma)\delta^j(h^{ji}\circ\sigma)] \\ + \delta^j(F\circ Y_\ell)g(\sigma)h^{ji}(\sigma)\}\end{aligned} \tag{4-12}$$

defines a function $\Delta_F^i\in C_p^{q-1}(E_{\ell+1})$, and we set $\Delta_F = (\Delta_F^i)_{1\leq i\leq d}$.

Now, if $\Psi^{ij}=g(\sigma)h^{ji}(\sigma)F(Y_\ell)$, we have $\Psi^{ij}= \det(\sigma)^2\ (\sigma^{-1})^{ji}\ F(Y_\ell)$, and 4-5 and 4-12 yield $\sum_{j\leq d}[\Psi^{ij}\gamma^j + \delta^j(\Psi^{ij})] = \Delta_F^i(Y_{\ell+1})$. Therefore applying 4-6 for Ψ^{ij} and summing on j gives

$$E[\frac{\partial}{\partial x_i}f(\Phi)\det(\sigma)^2 F(Y_\ell)] = E[\, f(\Phi)\Delta_F^i(Y_{\ell+1})]. \tag{4-13}$$

This is our basic integration-by-parts formula, which we can now iterate.

4-14 LEMMA: *Set* $Q=\det(\sigma)$. *Let* $n\in\mathbb{N}^*$, $\ell\in\mathbb{N}$, $i\in\mathbb{N}$, $s\geq 0$, $\varepsilon>0$, *and assume that* $C_{n+\ell-1}\subset H$ *and that* $Q^{-1}\in L^{2n+s+\varepsilon}$ (which implicitely means that $Q\neq 0$ a.s.). *For each* $F\in C_p^{n+i+1}(E_\ell)$ *there is a function* $\Delta_{F,n,s}$ *whose components belong to* $C_p^{i+1}(E_{\ell+n})$, *such that for all* $f\in C_p^{n+1}(R^d)$:

$$E[D^n_{x^n} f(\Phi)Q^{-s}F(Y_\ell)] = E[f(\Phi)Q^{-s-2n}\Delta_{F,n,s}(Y_{\ell+n})] \quad (4\text{-}15)$$

Proof. a) We firstly prove the claims for n=1. Let $w\geq 0$, $m\in\mathbb{N}$, $j\in\mathbb{N}$ and assume that $C_m\subset H$ and that $Q^{-1}\in L^{2+w+\varepsilon}$ for some $\varepsilon>0$. Let $F\in C_p^{j+1}(E_m)$, and $\phi\in C_b^\infty(R)$ be such that

$$0\leq\phi\leq 1,\ \phi(z)=0 \text{ for } |z|\leq\tfrac{1}{2},\ \phi(z)=1 \text{ for } |z|\geq 1.$$

Let $q\in\mathbb{N}^*$. We shall apply 4-13 to the function $F_q\in C_p^{i+1}(E_m)$ which meets $F_q(Y_m)=F(Y_m)g(\sigma)^{-w-2}\phi[qg(\sigma)]$: if ϕ' is the first derivative of ϕ, 4-5 yields

$$\begin{aligned}\delta(F_q\circ Y_m) &= \delta(F\circ Y_m)g(\sigma)^{-w-2}\phi[qg(\sigma)]\\ &\quad -(w+2)F(Y_m)\delta(g\circ\sigma)g(\sigma)^{-w-3}\phi[qg(\sigma)]\\ &\quad + F(Y_m)\delta(g\circ\sigma)g(\sigma)^{-w-2}\, q\,\phi'[qg(\sigma)],\end{aligned}$$

and so 4-12 yields

$$\begin{aligned}\Delta^i_{F_q}(Y_{m+1}) = \textstyle\sum_{j\leq d} \{&Q^{-w-2}\phi(qQ)[F(Y_m)h^{ji}(\sigma)\gamma^j g(\sigma)\\ &+ F(Y_m)\delta^j(h^{ji}\circ\sigma)g(\sigma) + \delta^j(F\circ Y_m)h^{ji}(\sigma)g(\sigma)\\ &-(w+1)F(Y_m)\delta^j(g\circ\sigma)h^{ji}(\sigma)]\\ &+ F(Y_m)\delta^j(g\circ\sigma)h^{ji}(\sigma)\, q\,\phi'(qQ)Q^{-w-1}\}.\end{aligned}$$

Let $q\uparrow\infty$. The properties of ϕ and the assumption

$Q^{-1}\in L^{2+w+\varepsilon}$ imply that $Q^{-w-2}\phi(qQ)\to Q^{-w-2}$ and that $q\phi'(qQ)Q^{-w-1}\to 0$ in $L^{1+\varepsilon}$, while all other terms in the previous formula are in $\cap_{p<\infty}L^p$. Hence $\Delta^i_{F_q}(Y_{m+1})$ converges in L^1 to the random variable $\Delta_{F,1,w}(Y_{m+1})Q^{-w-2}$, where $\Delta^i_{F,1,w}\in C^j_p(E_{m+1})$ satisfies

$$\begin{aligned}\Delta^i_{F,1,w}(Y_{m+1}) &= \textstyle\sum_{j\leq d}\{[F(Y_m)h^{ji}(\sigma)\gamma^j \\ &\quad + F(Y_m)\delta^j(h^{ji}\circ\sigma) + \delta^j(F\circ Y_m)h^{ji}(\sigma)]g(\sigma) \qquad (4\text{-}16)\\ &\quad - (w+1)F(Y_m)\delta^j(g\circ\sigma)h^{ji}(\sigma)\}.\end{aligned}$$

Similarly, if $f\in C^2_p(R^d)$, then $D_xf(\Phi)Q^{-w}\phi(qQ)F(Y_m)$ converges in L^1 to $D_xf(\Phi)Q^{-w}F(Y_m)$ as $q\uparrow\infty$. Since by 4-13

$$\begin{aligned}E[D_xf(\Phi)Q^{-w}\phi(qQ)F(Y_m)] &= E[D_xf(\Phi)Q^2F_q(Y_m)]\\ &= E[f(\Phi)\Delta_{F_q}(Y_m)],\end{aligned}$$

letting $q\uparrow\infty$ yields

$$E[D_xf(\Phi)Q^{-w}F(Y_m)] = E[f(\Phi)Q^{-w-2}\Delta_{F,1,w}(Y_{m+1})]. \qquad (4\text{-}17)$$

b) Now, this formula is easily iterated. Let n,ℓ, i,s,F,f be as in the statement of our lemma. We start applying 4-17 with $D^{m-1}_{x^{m-1}}f$ instead of f, and $w=s$, $m=\ell$, $j=n+i+1$, $F_0=F$. Then we apply 4-17 with $D^{m-2}_{x^{m-2}}f$ and $w=s+2$, $m=\ell+1$, $j=n+i$, $F_1=\Delta_{F_0,1,s}$. Then we apply 4-17 with $D^{m-3}_{x^{m-3}}f$ and $w=s+4$, $m=\ell+2$, $j=n+i-1$, $F_2=\Delta_{F_1,1,s+2}$, and so on... we end up with a function $F_n=\Delta_{F_{n-1},1,s+2n-2}$, which we call $\Delta_{F,n,s}$ and whose components belong to $C^{i+1}_p(E_{\ell+n})$, and which satisfies 4-15.

4-18 <u>REMARK</u>: The formula giving $\Delta_{F,n,s}$, obtained by iteration of 4-16, is quite complicated. However, an induction on the number of iterations easily gives the following property: if

$$|D^k_{x^k} F(y_\ell)| \leq \zeta(1+|y_\ell|^\theta), \quad k=0,1,..,n+i+1$$

(y_ℓ is the generic point of E_ℓ), then there are constants ζ', θ' depending on ζ,θ,ℓ,n,i,s (but not on F) such that

$$|D^k_{x^k}(\Delta_{F,n,s})(y_{\ell+m})| \leq \zeta'(1+|y_{\ell+m}|^{\theta'}), \; k=0,1,\cdots,i+1.$$

Now we easily deduce the following criterion:

4-19 THEOREM: *Assume 4-4 and let* $Q=\det(\sigma)$. *Define* C_r *by 4-10 with* $C_0=\{\gamma^i,\sigma^{ij}: i,j\leq d\}$. *In order that* Φ *admits a density of class* C^r, *it suffices that one of the following conditions holds:*

(i) $C_{r+d}\subset H$ *and* $Q^{-1}\in L^{2r+2d+2+\varepsilon}$ *for some* $\varepsilon>0$,

(ii) $C_r\subset H$ *and* $Q^{-1}\in L^{2d(r+1)+\varepsilon}$ *for some* $\varepsilon>0$.

Proof. Assume (i) first. The assumptions allow to apply 4-15 with $s=0$, $\ell=0$, $F=1$, for all $n\leq r+d+1$. Thus 4-2 holds for all $|\alpha|\leq r+d+1$, with $C:=\sup_{0\leq n\leq r+d+1} E(|Q|^{-2n}|\Delta_{1,n,0}(Y_n)|)$ and the claim follows from 4-1-a-(ii).

Secondly, assume (ii). We apply 4-15 with $s=0$, $\ell=0$, $F=1$ for all $n\leq r+1$: we get 4-3 for each multi-index α with $|\alpha|\leq r+1$, where Ψ_α is the relevant component of $Q^{-2|\alpha|}\Delta_{1,|\alpha|,0}(Y_{|\alpha|})$. Then $\Psi_\alpha\in L^{d+\varepsilon'}$ for some $\varepsilon'>0$, and the claim follows from 4-1-b.

The advantages of each one of (i) or (ii) above are obvious: more regularity, less integrability, needed in (i) than in (ii).

§4-c. JOINT SMOOTHNESS OF THE DENSITY

The present two subsections should be read only when needed, and only by somebody interested in the joint regularity of $p_t(x,y)$ (Theorems 2-28 and 2-29).

We suppose that $\{\Phi^x\}_{x\in\mathbb{R}^d}$ is a family of r.v., each one being d-dimensional. The following definitions are classical:

4-20 DEFINITIONS: a) The family $\{\Phi^x\}$ is F-*continuous* (resp. F-*differentiable*) if the map $x \rightsquigarrow \Phi^x$ is continuous (resp. Fréchet-differentiable) in L^p for all $p\in[1,\infty)$, at every point $x\in\mathbb{R}^d$. The derivative is denoted by $\nabla\Phi^x$.

b) The words "r times F-differentiable" or "r times F-continuously differentiable" are self-explaining. We denote by $\nabla^1\Phi^x=\nabla\Phi^x,\dots,\nabla^k\Phi^x,\dots$ the successive derivatives (and $\nabla^0\Phi^x=\Phi^x$).

For each $x\in\mathbb{R}^d$ let $(\sigma^x,\gamma^x,H^x,\delta^x)$ be an integration-by-parts setting for Φ^x. Suppose that $\{\Phi^x\}$ is j times F-differentiable; for each $q=0,1,\dots,j$ we consider a finite set of real r.v. $C_0^x(q)$, including all components of γ^x, σ^x, $\nabla^i\Phi^x$ for $0\leq i\leq q$. Then we define $C_r^x(q)$ by 4-10, starting with $C_0=C_0^x(q)$. As above, we call $Y_r^x(q)$ the multi-dimensional r.v. whose components are all the r.v. constituting the family $C_r^x(q)$.

4-21 THEOREM: *With the above notation, we assume that*

a) $\{\Phi^x\}$ *is* j *times F-differentiable.*

b) $C_{j-q}^x(q)\subset H^x$ *for* $1\leq q\leq j$, *and* $C_{j-1}^x(0)\subset H^x$.

c) $x \rightsquigarrow E(|Y_n^x(q)|^p)$ *is locally bounded for all* $p<\infty$ *and* $1\leq n+q\leq j+1$ *if* $q\geq 1$, *and* $1\leq n\leq j$ *if* $q=0$.

Furthermore, set $Q^x = \det(\sigma^x)$. *In order that* Φ^x *admit a density* $y \rightsquigarrow p(x,y)$ *on* $\mathbb{R}^d$ *for all* $x \in \mathbb{R}^d$, *and that* $p(x,y)$ *be of class* C^r *in* (x,y), *it is sufficient that one of the following conditions holds:*

(i) $j \geq r+2d+1$ *and* $x \rightsquigarrow E(|Q^x|^{-2r-4d-2-\varepsilon})$ *is locally bounded for some* $\varepsilon > 0$, *or*

(ii) $j \geq r+1$ *and* $x \rightsquigarrow E(|Q^x|^{-4rd-4d-\varepsilon})$ *is locally bounded for some* $\varepsilon > 0$.

Proof. a) Let $j=r+2d+1$ in case (i), $j=r+1$ in case (ii). Let $f, \tilde{f} \in C_0^\infty(R^d)$. Let $n, q \in \mathbb{N}$ with $1 \leq n+q \leq j$. We have

$$\int D^q_{x^q} f(x)\ E(D^n_{x^n} f(\Phi^x))dx = (-1)^q \int \tilde{f}(x) D^q_{x^q}[E(D^n_{x^n} f(\Phi^x))]dx. \tag{4-22}$$

Assume first that $q \geq 1$. By (a), $D^n_{x^n} f(\Phi^x)$ is q times F-differentiable, and its qth derivative is of the form $\sum_{1 \leq i \leq q} D^{n+i}_{x^{n+i}} f(\Phi^x) g_{n,i}(\nabla\Phi^x, .., \nabla^{q-i+1}\Phi^x)$, where $g_{n,i}$ is a (vector-valued) polynomial. Recalling the definition of $Y_0^x(q-i+1)$, we can write $g_{n,i}(\nabla\Phi^x, ..)$ as $G_{n,i}(Y_0^x(q-i+1)$. Hence 4-22 equals

$$= (-1)^q \int \tilde{f}(x) E\{\sum_{1 \leq i \leq q} D^{n+i}_{x^{n+i}} f(\Phi^x) G_{n,i}(Y_0^x(q-i+1)\}dx$$

$$= (-1)^q \int \tilde{f}(x) E\{\sum_{1 \leq i \leq q} f(\Phi^x)(Q^x)^{-2(n+i)} \Delta_{G_{n,i}, n+i, 0}(Y^x_{n+i}(q-i+1))\}dx,$$

where we have used 4-15 with $s=0$, and (b) (recall that $n+q \leq j$, so $C^x_{n+i-1}(q-i+1) \subset H^x$ if $1 \leq i \leq q$), and the fact that $(Q^x)^{-1} \in L^{2j+\varepsilon}$. Therefore if

$$Z^x_{n,q} = (-1)^q \sum_{1 \leq i \leq q} (Q^x)^{-2(n+i)} \Delta_{G_{n,i}, n+i, 0}(Y^x_{n+i}(q-i+1))$$

we have proved that

$$\int E(D^{q+n}_{x^q y^n} F(x,\Phi^x))dx = \int E(F(x,\Phi^x) Z^x_{n,q})dx \tag{4-23}$$

for F of the form $F=\tilde{f}\otimes f$. If $q=0$, 4-15 immediately yields 4-23 for $F=f\otimes f$, with $Z^x_{n,0}=(Q^x)^{-2n}\Delta_{1,n,0}(Y^x_n(0))$. Both sides of 4-23 being linear if F and continuous in F for the usual topology on $C^\infty_o(R^d\times R^d)$, we deduce that 4-23 holds for all $F\in C^\infty_o(R^d\times R^d)$.

b) Let A be an open bounded subset of $\mathbb{R}^d$. We will apply Lemma 4-1 after replacing P by $\overline{P}_A(dx,d\omega)=1_A(x)dx\times P(d\omega)$, and Φ by $(x,\omega)\rightsquigarrow\overline{\Phi}(x,\omega)=(x,\Phi^x(\omega))$ and d by $\overline{d}=2d$. If α is a multi-index with $|\alpha|=n+q\leq j$ and if $\overline{\Psi}_\alpha$ is the relevant component of $(x,\omega)\rightsquigarrow Z^x_{n,q}(\omega)$ we deduce from 4-23 that

$$\overline{E}_A(D^\alpha F(\overline{\Phi})) = \overline{E}_A(F(\overline{\Phi})\overline{\Psi}_\alpha) \quad \text{for } F\in C^\infty_o(R^d\times R^d). \tag{4-24}$$

Now, we recall that $\Delta_{F,r,0}$ is a function with polynomial growth for $F=G_{n,i}$ or $F=1$. Then we easily deduce from the assumption (c) and from (i) (resp. (ii)) that $\overline{\Psi}_\alpha\in L^1(\overline{P}_A)$ (resp. $\overline{\Phi}_\alpha\in L^{2d+\varepsilon'}(\overline{P}_A)$ for some $\varepsilon'>0$). Therefore Lemma 4-1 yields that the law of $\overline{\Phi}$ on $\mathbb{R}^d\times\mathbb{R}^d$, under $\overline{P}_A$, has a density, say $p_A(x,y)$, which is of class C^r. Since $p_A(x,.)$ is clearly the density of Φ^x for almost all x in A and since A is arbitrarily large, we are finished (observe that the F-continuity of Φ^x and the continuity of $x\rightsquigarrow p_A(x,y)$ imply that $p_A(x,.)$ is indeed the density of Φ^x for all $x\in A$).

§4-d. MORE ON JOINT SMOOTHNESS

As the reader has undoubtedly noticed, the previous subsection will be applied to $\Phi^x=X^x_t$ for a given t, where X^x is the solution to 1-4. Here we want to study the joint smoothness in (t,x,y) of the density $p_t(x,y)$ of

X_t^x. However, in order to avoid introducing now the necessary assumptions on the coefficients of 1-4, we will stick to our abstract setting.

For each $t\in[t_0,t_1]$ we consider a family $\{\Phi_t^x\}_{x\in\mathbb{R}^d}$ of d-dimensional r.v. with their integration-by-parts settings $(\sigma_t^x,\gamma_t^x,H_t^x,\delta_t^x)$. The sets introduced in §4-c are now denoted by $C_{t,r}^x(q)$, and we also write $Y_{t,r}^x(q)$ for the multi-dimensional r.v. whose components constitute $C_{t,r}^x(q)$. Finally, we assume the following ad-hoc hypothesis, for which we need some terminology.

4-25 A family $(f_u)_{u\in U}$ of functions $\mathbb{R}^d \to \mathbb{R}^{d'}$ is said k-DUPG (for: k times Differentiable with Uniform Polynomial Growth) if each f_u is k times differentiable, and $|D_{x^\ell}^\ell f_u(x)|\leq\zeta(1+|x|^\theta)$ for $0\leq\ell\leq k$, where ζ,θ are constants independent of u.

4-26 HYPOTHESIS: *Let* $j\in\mathbb{N}$ *with* $j\geq 2$. *There exist:*

(i) a finite measure m *on a measurable space* $(U,\underline{\underline{U}})$;

(ii) two families of functions $(A_u^1)_{u\in U}$: $\mathbb{R}^d \to \mathbb{R}^d$ *and* $(A_u^2)_{u\in U}$: $\mathbb{R}^d \to \mathbb{R}^d\otimes\mathbb{R}^d$, *with* $(A_u^i)_{u\in U}$ *being* $(j+i-3)$-*DUPG;*

(iii) a family $(\phi_u)_{u\in U}$ *of functions:* $\mathbb{R}^d \to \mathbb{R}^d$ *which is* j-*DUPG, with*

$$|\det(D_x\phi_u(x))| \geq \zeta > 0, \text{ all } u\in U, x\in\mathbb{R}^d \tag{4-27}$$

Moreover, if $f\in C_p^2(R^d)$ *and if*

$$Lf(x) = \sum_{i=1}^{2} \int m(du) \langle A_u^i(x),(D_{x^i}^i f)\circ\phi_u(x)\rangle \tag{4-28}$$

(here, $\langle .,.\rangle$ is the scalar product on $(\mathbb{R}^d)^{\otimes i}$), *then* $t \rightsquigarrow E(f(\Phi_t^x))$ *is differentiable on* $[t_0,t_1]$, *with deriva-*

tive $E(Lf(\Phi_t^x))$.

We shall see later that the generator L' of X^x (see 1-3) can be written as 4-28. The form 4-28 easily allows for iterating the operator L, as seen in the following lemma:

4-29 LEMMA: *Assume 4-26 for some* $j\geq 2$. *Let* $n\in\mathbb{N}^*$ *with* $n\leq j/2$. *For all* $i=1,\ldots,2n$, *there is a family* $(B^{n,i}_{u_1,\ldots,u_n})_{u_i\in U}$ *of functions*: $\mathbb{R}^d\to(\mathbb{R}^d)^{\otimes i}$ *which is* $(j+i-2n-1)$-DUPG, *and such that for* $f\in C_p^{2n}(\mathbb{R}^d)$,

$$L^n f(x) = \sum_{i=1}^{2n}\int\ldots\int m(du_1)\ldots m(du_n) \langle B^{n,i}_{u_1,\ldots,u_n}(x),(D^i_{x^i}f)\circ\phi^n_{u_1,\ldots,u_n}(x)\rangle, \tag{4-30}$$

where $\phi^1_u=\phi_u$, $\phi^{n+1}_{u_1,\ldots,u_{n+1}} = \phi^n_{u_1,\ldots,u_n}\circ\phi_{u_{n+1}}$.

Proof. For $n=1$ the claim reduces to 4-26, with $B^{1,i}_u=A^i_u$. Suppose that the claim holds for some n, with $n+1\leq j/2$. Let $f\in C_p^{2n+2}(\mathbb{R}^d)$; then $L^n f$, as given by 4-30, is of class C_p^2 and we can apply L once more. Due to 4-28, it is clear that $L^{n+1}f$ is again of the form 4-30, with $2n+2$ terms, each one being an $(n+1)$-uple integral. Moreover, the contribution of the ith term of 4-30 to $L^{n+1}f$ is:

- for $B^{n+1,i}_{(..)u}$: $A^1_u(D_x B^{n,i})\circ\phi_u + A^2_u(D^2_{x^2}B^{n,i})\circ\phi_u$;
- for $B^{n+1,i+1}_{(..)u}$: $A^1_u(B^{n,i}(D_x\phi^n))\circ\phi_u + A^2_u(B^{n,i}(D^2_{x^2}\phi^n))\circ\phi_u + 2A^2_u[(D_xB^{n,i})(D_x\phi^n)]\circ\phi_u$;
- for $B^{n+1,i+2}_{(..)u}$: $A^2_u[B^{n,i}(D_x\phi^n)(D_x\phi^n)]\circ\phi_u$.

Then we easily deduce from the properties of $A^i_u, B^{n,i}_{..}, \phi_u$

and $\phi^n_{u_1,\dots,u_n}$ (which is also (j-1)-DUPG) that $B^{n+1,i}_{\dots}$ is $(j+1-2n-3)$-DUPG.

4-31 THEOREM: *Let* $j\in\mathbb{N}^*$ *be such that, with the notation introduced before 4-25:*

a) for $t\in(t_0,t_1)$, $\{\Phi^x_t\}_{x\in\mathbb{R}^d}$ *is* j *times F-differentiable;*

b) for $t\in(t_0,t_1)$, $C^x_{t,j-q}(q)\subset H^x_t$ *if* $0\leq q\leq j$;

c) for every bounded set A *we have for all* $p<\infty$, $0\leq n+q\leq j+1$: $\sup_{x\in A,t_0<t<t_1} E(|Y^x_{t,n}(q)|^p)<\infty$;

d) 4-26 holds for $j+2$.

Furthermore, let $Q^x_t=\det(\sigma^x_t)$. *In order that* Φ^x_t *admit a density* $y\leadsto p_t(x,y)$ *on* $\mathbb{R}^d$ *for all* $x\in\mathbb{R}^d$, $t\in(t_0,t_1)$, *which of class* C^r *in* (t,x,y) *on* $(t_0,t_1)\times\mathbb{R}^d\times\mathbb{R}^d$, *it is sufficient that one the following two conditions holds:*

(i) $j\geq 2r+4d+3$ *and, for every bounded set* A *there is* $\varepsilon>0$ *with* $\sup_{x\in A,t_0<t<t_1} E(|Q^x_t|^{-4r-8d-8-\varepsilon})<\infty$, *or*

(ii) $j\geq 2r+1$ *and, for every bounded set* A *there is* $\varepsilon>0$ *with* $\sup_{x\in A,t_0<t<t_1} E(|Q^x_t|^{-4(r+1)(2d+1)-\varepsilon})<\infty$, *and* $\phi_u(x)\equiv x$ *in 4-26.*

The restriction $\phi_u(x)\equiv x$ in (ii) is quite a serious one: applied to the generator L' given by 1-3, it means that $K=0$ (which corresponds to a process X^x which is continuous).

Proof. a) We set $\tilde{r}=r+2d+2$ in case (i), $\tilde{r}=r+1$ in case (ii), so that $j\geq 2\tilde{r}-1$ and that for every bounded set A there is $\varepsilon>0$ with

$$\sup_{x\in A,t_0<t<t_1} E(|Q^x_t|^{-4\tilde{r}-\varepsilon}) < \infty. \tag{4-32}$$

For each $\ell \geq 1$ we set $(U^\ell, \underline{\underline{U}}^\ell, m^\ell) = (U, \underline{\underline{U}}, m)^{\otimes \ell}$, and we also set $(U^0, \underline{\underline{U}}^0, m^0) = (\{0\}, \mathcal{P}(\{0\}), \varepsilon_0)$; set $\phi_0^0(x) = x$ and $B_0^{0,0}(x) = 1$ and $B_v^{n,0}(x) = 0$ for $n \geq 1$, $v \in U^n$. Thus in view of 4-30, and with the convention $L^0 f = f$, we have

$$L^n f(x) = \sum_{i=0}^{2n} \int m^n(dv) \langle B_v^{n,i}(x), (D^i_{x^i} f) \circ \phi_v^n(x) \rangle, \quad \text{for} \quad 0 \leq n \leq \tilde{r}. \tag{4-33}$$

b) Let $k, q, n \in \mathbb{N}$ be fixed, with $1 \leq k+q+n \leq \tilde{r}$. Let $\tilde{f} \in C_0^\infty((t_0, t_1))$, $f, \bar{f} \in C_0^\infty(\mathbb{R}^d)$. Using 4-26 and an integration by parts on (t_0, t_1), then 4-33, then an integration by parts on $\mathbb{R}^d$, we obtain (the various summations of components are straightforward, and left to the reader: there seems to be already enough indices coming from order of differentiation!):

$$\int_{t_0}^{t_1} dt \int D^k_{t^k} \tilde{f}(t) \; D^q_{x^q} \bar{f}(x) \; E(D^n_{x^n} f(\Phi_t^x)) dx \tag{4-34}$$

$$= (-1)^k \int_{t_0}^{t_1} dt \int \tilde{f}(t) \; D^q_{x^q} \bar{f}(x) \; E\{L^k(D^n_{x^n} f)(\Phi_t^x)\} dx$$

$$= (-1)^k \int_{t_0}^{t_1} dt \int dx \; \tilde{f}(t) \; D^q_{x^q} \bar{f}(x) \sum_{i=0}^{2k} \int_{U^k} m^k(dv) E(\langle B_v^{k,i}(\Phi_t^x), (D^{n+i}_{x^{n+i}} f) \circ \phi_v^k(\Phi_t^x) \rangle)$$

$$= (-1)^{k+q} \int_{t_0}^{t_1} dt \int dx \; \tilde{f}(t) \; \bar{f}(x) \sum_{i=0}^{2k} \int_{U^k} m^k(dv) \rho_i(t,x,v)$$

where

$$\rho_i(t,x,v) = D^q_{x^q} \{E(\langle B_v^{k,i}(\Phi_t^x), (D^{n+i}_{x^{n+i}} f) \circ \phi_v^k(\Phi_t^x) \rangle)\}.$$

Suppose first that $q \geq 1$. There are polynomials g_s (for $s \leq q$) such that for every function h of class C^q, then $\nabla^q h(\Phi_t^x) = \sum_{1 \leq s \leq q} D^s_{x^s} h(\Phi_t^x) g_s(Y_{t,0}^x(q-s+1))$. Hence

$$\nabla^q \{\langle B_v^{k,i}(\Phi_t^x), (D^{n+i}_{x^{n+i}} f) \circ \phi_v^k(\Phi_t^x) \rangle\}$$

$$= \sum_{\ell=0}^{q} D^{\ell}_{x^\ell} \{(D^{n+i}_{x^{n+i}} f)\circ\phi^k_v\}(\Phi^x_t)$$
$$\times \sum_{s=\ell\vee 1}^{q} D^{s-\ell}_{x^{s-\ell}} B^{k,i}_v(\Phi^x_t) g_s(Y^x_{t,0}(q-s+1))$$

$$= \sum_{\ell=0}^{q} \sum_{s=\ell\vee 1}^{q} D^{\ell}_{x^\ell} \{(D^{n+i}_{x^{n+i}} f)\circ\phi^k_v\}(\Phi^x_t) G^{i,s,\ell}_v(Y^x_{t,0}(q-s+1))$$

where the family $\{G^{i,s,\ell}_v\}_v$ is clearly $(2\tilde{r}+i-2k-s+\ell)$-DUPG. Of course $\rho_i(t,x,v)$ is the expected value of this expression. For ℓ,s as above, we have $C^x_{t,\ell-1}(q-s+1)\subset H^x_t$ and $\ell\leq\tilde{r}$ Hence we deduce from Lemma 4-14:

$$\rho_i(t,x,v) = \sum_{\ell=0}^{q} \sum_{s=\ell\vee 1}^{q} E\{(D^{n+i}_{x^{n+i}} f)\circ\phi^k_v(\Phi^x_t)\ (Q^x_t)^{-2\ell}$$
$$\hat{G}^{i,s,\ell}_v(Y^x_{t,\ell}(q-s+1))\},$$

where, in virtue of 4-18, the family $\{\hat{G}^{i,s,\ell}_v\}_v$ is $(2\tilde{r}+i-2k-s)$-DUPG. For $q=0$ we have the same expression, with only one term $\hat{G}^{i,0,0}_v(Y^x_{t,0}(0))=B^{k,i}_v(\Phi^x_t)$.

Next, for each $\theta=1,\ldots,j+2$ there are polynomials $h_{s,\theta}$ with

$$D^{\theta}_{x^\theta}(f\circ\phi^k_v) = \sum_{s=1}^{\theta} \{(D^s_{x^s} f)\circ\phi^k_v\} h_{s,\theta}(D_x\phi^k_v,\ldots,D^{\theta-s+1}_{x^{\theta-s+1}}\phi^k_v),$$

where for $s=\theta$, $h_{\theta,\theta}(\ldots)=[D_x\phi^k_v]^{\otimes\theta}$. Due to 4-27, we can thus invert the triangular system above, so that

$$(D^s_{x^s} f)\circ\phi^k_v = \sum_{\theta=1}^{s} \{D^\theta_{x^\theta}(f\circ\phi^k_v)\} h^{s,\theta}_v$$

where, due again to 4-26-(iii), the family $\{h^{s,\theta}_v\}_v$ is $(2\tilde{r}-s+\theta)$-DUPG. Hence

$$\rho_i(t,x,v) = \sum_{\ell=0}^{q} \sum_{s=\ell\vee 1}^{q} \sum_{\theta=1}^{n+i} E\{D^\theta_{x^\theta}(f\circ\phi^k_v)(\Phi^x_t)\ (Q^x_t)^{-2\ell}$$
$$\hat{G}^{i,s,\ell}_v(Y^x_{t,\ell}(q-s+1))\ h^{s,\theta}_v(\Phi^x_t)\}.$$

For θ,ℓ as above, we have $C^x_{t,\theta+\ell-1}(q-s+1)\subset H^x_t$ if $q\geq 1$, and $C^x_{t,\theta-1}(0)\subset H^x_t$ if $q=0$, while $E(|Q^x_t|^{-2\ell-2\theta-\varepsilon})<\infty$. Applying again Lemma 4-14 yields

$$\rho_i(t,x,v) = \sum_{\ell=0}^{q}\sum_{s=\ell\vee 1}^{q}\sum_{\theta=1}^{n+i} E\{f\circ\phi^k_v(\Phi^x_t)\,(Q^x_t)^{-2\ell-2\theta}\, H^{i,s,\ell,\theta}_v(Y^x_{t,\ell+\theta}(q-s+1))\}$$

where the family $\{H^{i,s,\ell,\theta}_v\}_v$ is (at least) (0)-DUPG (if $q=0$ there is only the sum over θ, and one reads $H^{i,0,0,\theta}_v(Y^x_{t,\theta}(0))$). Therefore if

$$Z^x_t(v) = (-1)^{k+q}\sum_{i=0}^{2k}\sum_{\ell=0}^{q}\sum_{s=\ell\vee 1}^{q}\sum_{\theta=1}^{n+i}(Q^x_t)^{-2\ell-2\theta}\, H^{i,s,\ell,\theta}_v(Y^x_{t,\ell+\theta}(q-s+1)) \tag{4-35}$$

we see that 4-34 equals

$$\int_{t_0}^{t_1} dt \int dx\ \tilde{f}(t)\ \bar{f}(x)\ E\{\sum_{i=0}^{2n}\int m^k(dv)\ Z^x_t(v)\ f\circ\phi^k_v(\Phi^x_t)\}.$$

In other words, we have

$$\int_{t_0}^{t_1} dt\int dx\ E\{D^{k+q+n}_{t^k x^q y^n}F(t,x,\Phi^x_t)\} = \int_{t_0}^{t_1} dt\int dx\ E\{\int m^k(dv)Z^x_t(v)F(t,x,\phi^k_v(\Phi^x_t))\}, \tag{4-36}$$

for all $F=\tilde{f}\otimes\bar{f}\otimes f$ and, by linearity and continuity it immediately extends to all $F\in C^\infty_o((t_0,t_1)\times R^d\times R^d)$.

c) Let A be an open bounded subset of $\mathbb{R}^d$. We will apply Lemma 4-1 to $\tilde{P}_A(dt,dx,d\omega)= 1_{(t_0,t_1)}(t)dt\times 1_A(x)dx\times P(d\omega)$, and $\tilde{\Phi}(t,x,\omega)=(t,x,\Phi^x_t(\omega))$, and $\tilde{d}=2d+1$. Then for $F\in C^\infty_o((t_0,t_1)\times R^d\times R^d)$, 4-36 reads

$$\tilde{E}_A(D^{k+q+n}_{t^k x^q y^n}F(\tilde{\Phi})) = \int_{t_0}^{t_1} dt\int_A dx\ E\{\int_{U^k} m^k(dv)Z^x_t(v)F(t,x,\Phi^x_t))\}. \tag{4-37}$$

Now, 4-32 and (c) and the (0)-DUPG property of $\{H_v^{i,s,\ell,\theta}\}_v$ and 4-35 imply that

$$\sup_{x\in A, v\in U^k, t_0<t<t_1} E(|Z_t^x(v)|) < \infty.$$

Therefore there is a constant $C_{k,q,n}$ such that

$$|\tilde{E}_A\{D^{k+q+n}_{t^k x^q y^n} F(\tilde{\Phi})\}| \leq C_{k,q,n} \|F\|_\infty .$$

This being true for all $k+q+n\leq\tilde{r}$ we deduce the result under hypothesis (i), from Lemma 4-1-a.

d) It remains to prove the result under (ii). This hypothesis clearly implies that

$$\sup_{x\in A, v\in U^k, t_0<t<t_1} E(|Z_t^x(v)|^{\tilde{d}+\varepsilon'}) < \infty$$

for some $\varepsilon'>0$ (recall that $\tilde{d}=2d+1$). Then if $|\alpha|=k+q+n$ and if Ψ_α is the relevant component of $(t,x,\omega) \rightsquigarrow \int m^k(dv) Z_t^x(v,\omega)$, using now $\phi_v^k(x)=x$, we deduce from 4-37 that

$$\tilde{E}_A(D^\alpha F(\tilde{\Phi})) = \tilde{E}_A(F(\tilde{\Phi})\Psi_\alpha)$$

and since $\Psi_\alpha\in L^{\tilde{d}+\varepsilon'}(\tilde{P}_A)$ the result follows from Lemma 4-1-b (same argument as at the end of the proof of 4-21).

Section 5: STABILITY FOR STOCHASTIC DIFFERENTIAL EQUATIONS

In this section, the basic ingredient is a filtered probability space $(\Omega, \underline{\underline{F}}, (\underline{\underline{F}}_t)_{t \in [0,T]}, P)$ endowed with

a) an m-dimensional Brownian motion $W=(W^i)_{i \leq m}$,

b) a Poisson random measure μ on $[0,T] \times E$, where $(E, \underline{\underline{E}})$ is some arbitrary measurable space. We suppose that the intensity measure of μ is $\nu(dt,dx) = dt \times G(dx)$, where G is a positive σ-finite measure on $(E, \underline{\underline{E}})$, and as usual $\tilde{\mu} = \mu - \nu$.

These assumptions implies in particular that W and μ are independent.

The L^p estimates needed to establish existence, uniqueness and stability or differentiability are all consequences of the following lemma:

5-1 LEMMA: *Let* $\eta \in \cap_{2 \leq p < \infty} L^p(E,G)$ *and* $d \in \mathbb{N}^*$. *For each* $p \in [2, \infty)$ *there is a constant* δ_p (depending on p, and also on T, η, m, d) *such that*

a) $E(|H*t|_t^{*p}) \leq \delta_p \int_0^t E(|H_s|^p) ds$ *if* H *is an* $\mathbb{R}^d$-*valued process;*

b) $E(|K*W|_t^{*p}) \leq \delta_p \int_0^t E(|K_s|^p) ds$ *if* K *is an* $\mathbb{R}^d \otimes \mathbb{R}^m$-*valued predictable process;*

c) $E(|U*\tilde{\mu}|_t^{*p}) \leq \delta_p \int_0^t E(L_s^p) ds$ *if* U *is an* $\mathbb{R}^d$-*valued* $\underline{\underline{P}} \otimes \underline{\underline{E}}$-*measurable function on* $\Omega \times [0,T] \times E$ *and* L *is a predictable process with* $|U(\omega,s,z)| \leq L_s(\omega) \eta(z)$.

5-2 REMARKS: 1) In particular, if the right-hand side of these inequalities is finite, the corresponding stochastic integrals on the left side exist.

2) On could actually get δ_p independent of the dimensions d and m.

3) This lemma is proved in [5] for p of the form $p=2^r$, $r\in\mathbb{N}^*$, and we really need it later only for such p's. However, it is true for all $p\in[2,\infty)$: this is trivial for (a), and a trivial consequence of Davis-Burkhölder-Gundy inequalities for (b); for (c) it needs an interpolation argument as the one given in [2] for example.

§5-a. GRADED STOCHASTIC EQUATIONS

We consider equations of the form

$$Y = H + A(Y_-)*t + B(Y_-)*W + C(Y_-)*\tilde{\mu} \tag{5-3}$$

where H is an $\mathbb{R}^d$-valued càdlàg process (the "initial condition") and

$$\left.\begin{array}{ll} A: & \mathbb{R}^d\times\Omega\times[0,T] \to \mathbb{R}^d \quad \text{is } \underline{\underline{\mathbb{R}}}^d\otimes\underline{\underline{P}}\text{-measurable} \\ B: & \mathbb{R}^d\times\Omega\times[0,T] \to \mathbb{R}^d\otimes\mathbb{R}^m \text{ is } \underline{\underline{\mathbb{R}}}^d\otimes\underline{\underline{P}}\text{-measurable} \\ C: & \mathbb{R}^d\times\Omega\times[0,T]\times E \to \mathbb{R}^d \text{ is } \underline{\underline{\mathbb{R}}}^d\otimes\underline{\underline{P}}\otimes\underline{\underline{E}}\text{-measurable.} \end{array}\right\} \tag{5-4}$$

We shall need coefficients not globally Lipschitz, but which have a "Lipschitz lower triangular structure" already encountered by Stroock [26] (for the very same reasons). This motivates the following:

5-5 DEFINITION: a) A *grading* of $\mathbb{R}^d$ is a decomposition $\mathbb{R}^d = \mathbb{R}^{d_1}\times\ldots\times\mathbb{R}^{d_q}$ with $d_1+\ldots+d_q=d$. The coordinates of a point in $\mathbb{R}^d$ are always arranged in an increasing order along the subspaces $\mathbb{R}^{d_i}$, and we set $M_0=0$ and $M_\ell=d_1+..+d_\ell$ for $1\leq\ell\leq q$.

b) We say that the coefficients (A,B,C) are *graded* according to the grading $\mathbb{R}^d = \mathbb{R}^{d_1}\times...\times\mathbb{R}^{d_q}$ if $A^i(y,\omega,t)$ and $B^{ij}(y,\omega,t)$ and $C^i(y,\omega,t,z)$ depend upon y only through the coordinates $(y^k)_{1\leq k\leq M_r}$ when $M_{r-1}<i\leq M_r$.

5-6 REMARK: Call Π_ℓ the orthogonal projection on $\mathbb{R}^{d_1}\times...\times\mathbb{R}^{d_\ell}$, in $\mathbb{R}^d$ ($\Pi_0=0$). Then (b) above can be written

$$\Pi_\ell A(y,.) = \Pi_\ell A(\Pi_\ell y,.), \quad \ell=1,...,q \tag{5-7}$$

and the same for $B^{\cdot k}$ and C. Of course, since 4-3 is "coordinate-free" one could equivalently define a grading as being an arbitrary increasing sequence $(\Pi_0=0,\Pi_1,...,\Pi_q=I)$ of orthogonal projections. Then A would be *graded* if 5-7 hold.

The inequalities 5-1 lead us to define the following norm, for processes:

$$|||Z|||_p = \{\int_0^T E(|Z_s|^p)ds\}^{1/p} \tag{5-8}$$

5-9 HYPOTHESES ON THE COEFFICIENTS: (A,B,C) *are graded according to* $\mathbb{R}^{d_1}\times...\times\mathbb{R}^{d_q}$. *Moreover let* $\eta\in\cap_{2\leq p<\infty}L^p(E,G)$, *and if* F *is any one of the functions* $F(y,\omega,t,z) = A(y,\omega,t)$, *or* $B^{\cdot k}(y,\omega,t)$, *or* $\frac{C(y,\omega,t,z)}{\eta(z)}$, *then* F *is differentiable in* y *on* $\mathbb{R}^d$ *and*

(i) $|F(0,\omega,t,z)| \leq Z_t(\omega)$

(ii) $|D_yF(y,\omega,t,z)| \leq \hat{Z}_t(\omega)(1+|y|^\theta)$

(iii) $|\frac{\partial}{\partial y_j} F^i(y,\omega,t,z)| \leq \zeta$ if $M_{r-1} \leq i,j \leq M_r$ for some $r \leq q$

where $\zeta, \theta \geq 0$ *are constants, and* $Z, \hat{Z}$ *are predictable processes such that* $|||Z|||_p$ *and* $|||\hat{Z}|||_p$ *are finite for all* $p \in [1,\infty)$.

5-10 THEOREM: *Assume 5-9 and also that* $|H|_T^* \in \cap_{p<\infty} L^p$. *Then 5-3 admits a unique solution* Y (up to evanescent sets), *and for every* $p \in [1,\infty)$ *there are constants* c_p *and* γ_p *depending only upon* $(\zeta, \theta, \{|||\hat{Z}|||_r\}_{r \in [1,\infty)})$, *such that*

$$\|Y_T^*\|_{L^p} \leq c_p(\|H_T^*\|_{L^{\gamma_p}} + |||Z|||_{\gamma_p}). \tag{5-11}$$

Proof. a) If the grading is trivial (i.e. q=1 in 5-5) the coefficients are globally Lipschitz; existence and uniqueness and also $Y_T^* \in \cap_{p<\infty} L^p$ are then well known (see for example [5,A6]; adding a random initial condition H satisfying $|H|_T^* \in \cap_{p<\infty} L^p$ changes nothing).

b) Consider the general case. We use notation Π_ℓ of 5-6 and we set $Q_\ell = \Pi_\ell - \Pi_{\ell-1}$. If Y satisfies 5-3, then by 5-7

$$\Pi_\ell Y = \Pi_\ell H + \Pi_\ell A(\Pi_\ell Y_-) * t + \Pi_\ell B(\Pi_\ell Y_-) * W + \Pi_\ell C(\Pi_\ell Y_-) * \tilde{\mu}. \tag{5-12}$$

Due to (a), 5-12 for $\ell=1$ has a unique solution $\Pi_1 Y$, and $|\Pi_1 Y|_T^* \in \cap_{p<\infty} L^p$.

We proceed by induction on ℓ. Suppose that for some $\ell \geq 1$, 5-12 has a unique solution $\Pi_\ell Y$ and that $|\Pi_\ell Y|_T^* \in \cap_{p<\infty} L^p$. Then X is a solution to 5-12 for $\ell+1$ if and only if $\Pi_\ell X = \Pi_\ell Y$ and

$$Q_{\ell+1}X = Q_{\ell+1}H + \tilde{A}(Q_{\ell+1}X_-)*t + \tilde{B}(Q_{\ell+1}X_-)*W + \tilde{C}(Q_{\ell+1}X_-)*\tilde{\mu}, \tag{5-13}$$

where $\tilde{A}(x,\omega,t)=Q_{\ell+1}A((\Pi_\ell Y_{t-}(\omega)+Q_{\ell+1}x),\omega,t)$ and similarly for $\tilde{B}$ and $\tilde{C}$. From 5-9-(iii) we deduce that $(\tilde{A},\tilde{B},\tilde{C})$ are globally Lipschitz, and they also meet 5-9-(i) with $\tilde{Z}_t = Z_t+\hat{Z}_t(|\Pi_\ell Y_{t-}|+|\Pi_\ell Y_{t-}|^{\theta+1})$, which clearly satisfies $|||\tilde{Z}|||_p<\infty$ for all $p\in[1,\infty)$ (recall that $|\Pi_\ell Y|^*_T\in\cap_{p<\infty}L^p$). Applying (a) again yields that 5-13 has a unique solution $Q_{\ell+1}X$, with $|Q_{\ell+1}X|^*_T\in\cap_{p<\infty}L^p$. Thus our induction hypothesis obtains for $\ell+1$: hence 5-3 has a unique solution Y, and $Y^*_T\in\cap_{p<\infty}L^p$.

c) It remains to prove the estimate 5-11. Here again we suppose first that the grading is trivial. So $|A(Y_-)|\leq Z+\zeta|Y_-|$ and similarly for B and for C/η. Thus if $b_p(t)=E(Y^{*p}_t)$, Lemma 5-1 yields a constant δ_p (not depending on H,A,B,C) such that

$$\begin{aligned} b_p(t) &\leq \delta_p\{E(H^{*p}_t) + \int_0^t E(Z^p_s)ds + \zeta^p\int_0^t E(Y^{*p}_s)ds\} \\ &\leq \delta_p\{\|H^*_T\|^p_{L^p} + |||Z|||^p_p + \zeta^p\int_0^t b_p(s)ds\}. \end{aligned} \tag{5-14}$$

We know that $b_p(t)<\infty$, so Gronwall's Lemma yields

$$b_p(t) \leq \delta_p\, e^{(\zeta^p T\delta_p)}\{\|H^*_T\|^p_{L^p} + |||Z|||^p_p\}.$$

In other words, there is a constant δ'_p (depending on ζ) such that

$$\|Y^*_T\|_{L^p} \leq \delta'_p\{\|H^*_T\|_{L^p} + |||Z|||_p\}. \tag{5-15}$$

Now, the general case: set $a_p(\ell)=\|\Pi_\ell Y^*_T\|_{L^p}$, and $f_p(\ell+1)=\|Q_{\ell+1}Y^*_T\|_{L^p}$. We have seen that $Q_{\ell+1}Y$ satisfies 5-13, whose coefficients meet 5-9 with

$\tilde{Z} = Z+\hat{Z}(|\Pi_\ell Y_-|+|\Pi_\ell Y_-|^{\theta+1})$ instead of Z. Then 5-15 yields

$$\begin{aligned} a_p(\ell+1) &\leq a_p(\ell) + f_p(\ell+1) \\ &\leq a_p(\ell) + \delta'_p\{\|H^*_T\|_{L^p} + |||Z|||_p \\ &\quad + |||\hat{Z}|||_{2p}a_{2p}(\ell) + |||\hat{Z}|||_{2p}a_{2p(\theta+1)}(\ell)\}. \end{aligned} \tag{5-16}$$

Moreover $a_p(0)=0$. Then an induction on ℓ allows to easily deduce 5-11 from 5-16.

5-17 COROLLARY: *Let* (H;A,B,C) *and* (H';A',B',C') *satisfy the assumptions of 5-10, with the same processes* $Z,\hat{Z}$ *and the same function* η *in 5-9, and call* Y *and* Y' *the corresponding solutions. Suppose also that for any one of the functions* F *in 5-9, and the corresponding* F' (associated with A',B',C'),

$$|F(y,\omega,t,z)-F'(y,\omega,t,z)| \leq \tilde{Z}_t(\omega)(1+|y|^\theta) \tag{5-18}$$

where $\tilde{Z}$ *is predictable. Then for every* $p\in[1,\infty)$ *there are constants* c'_p *and* γ'_p *depending upon* $(\zeta,\theta,\{|||\hat{Z}|||_r\}_{r\in[1,\infty)},\{\|Y^*_T\|_{L^r}\}_{r\in[1,\infty)})$, *such that*

$$\|(Y-Y')^*_T\|_{L^p} \leq c'_p(\|(H-H')^*_T\|_{L^{\gamma'_p}} + |||\tilde{Z}|||_{\gamma'_p}). \tag{5-19}$$

Proof. $Y''=Y-Y'$ is solution to an equation 5-3 with

$$\left.\begin{aligned} H''&=H-H'+[A(Y_-)-A'(Y_-)]*t \\ &\quad + [B(Y_-)-B'(Y_-)]*W + [C(Y_-)-C'(Y_-)]*\tilde{\mu} \\ A''(y,\omega,t) &= A'(Y_{t-}(\omega),\omega,t)-A'(Y_{t-}(\omega)-y,\omega,t) \end{aligned}\right\} \tag{5-20}$$

and the same for B",C". These coefficients satisfy 5-9 with $Z''_t=0$ and $\hat{Z}''_t=2^{\theta-1}\hat{Z}_t(1+|Y|^{*\theta}_{t-})$, so

$|||\hat{Z}''|||_p \le c(p,\theta)(|||\hat{Z}|||_{2p} + \|Y_T^*\|_{L^{2p}})$ for a constant $c(p,\theta)$. Hence 5-11 yields two constants c_p, γ_p depending only upon $(\zeta,\theta,\{|||\hat{Z}|||_r\}_{r\in[1,\infty)},\{\|Y_T^*\|_{L^r}\}_{r\in[1,\infty)})$, such that $\|(Y-Y')_T^*\|_{L^p} \le c_p \|H_T^*\|_{L^{\gamma_p}}$. Moreover Lemma 5-1 applied to 5-20, and 5-18, yield a constant δ_p such that

$$E(H''^{*\,\gamma_p}_T) \le \delta_p\{E(|H-H'|_T^{*\,\gamma_p}) + \int_0^T E(\tilde{Z}_t^{\gamma_p}(1+|Y|_T^{*\,\theta\gamma_p}))dt\}$$

$$\le \delta_p\{E(|H-H'|_T^{*\,\gamma_p}) + \sqrt{T}|||\tilde{Z}|||_{2\gamma_p}^{\gamma_p} + \sqrt{T}|||\tilde{Z}|||_{2\gamma_p}^{\gamma_p} \|Y_T^*\|_{L^{2\theta\gamma_p}}^{\gamma_p}\}.$$

Then we get 5-19, with $\gamma'_p = 2\gamma_p$ and $c'_p = \delta_p(1+T^{1/2\gamma_p}\|Y_T^*\|_{L^{2\theta\gamma_p}})$.

§5-b. DIFFERENTIABILITY

We have already encountered F-differentiability in 4-20 for random variables. For processes it goes as follows (Λ is a neighbourhood of 0 in $\mathbb{R}^n$).

5-21 DEFINITIONS: a) A family $(H^\lambda)_{\lambda\in\Lambda}$ of processes is called *F-differentiable at* 0, *with derivative* $\partial H = (\partial H^i)_{i\le n}$ (the derivative is also a process) if

- $|H^\lambda|_T^* \in \cap_{p<\infty} L^p$ for all $\lambda\in\Lambda$,
- $\|(H^\lambda - H^0 - \partial H.\lambda)_T^*\|_{L^p} = o(|\lambda|)$ as $\lambda\to 0$, for all $p\in[1,\infty)$.

b) If $(U^\lambda)_{\lambda\in\Lambda}$ is a family of functions on $\Omega\times[0,T]\times E$, the definition is identical, provided we replace $|H^\lambda|_T^*$ by $\sup_{s,z}|U^\lambda(\omega,s,z)|$ and $|H^\lambda - H^0 - \partial H.\lambda|_T^*$ by $|Z^\lambda|_T^*$, where Z^λ is a *predictable process* such that $|U^\lambda(s,z)-U^0(s,z)-\partial U(s,z).\lambda| \le Z_s^\lambda$: the derivative ∂U is then a function on $\Omega\times[0,T]\times E$ as well (the predictability of Z^λ is an ad-hoc hypothesis, aimed to avoid repeating over and over the same assumption).

(According to our standing conventions, the derivative is a row vector-valued process when H^λ is scalar-valued, or a tensor-valued process with one more covariant index when H^λ is a tensor).

Now, we consider a family of equations of the form 5-3:

$$Y^\lambda = H^\lambda + A^\lambda(Y^\lambda_-)*t + B^\lambda(Y^\lambda_-)*W + C^\lambda(Y^\lambda_-)*\tilde{\mu} \qquad (5\text{-}22)$$

indexed by a bounded neighbourhood Λ of 0 in $\mathbb{R}^n$.

5-23 <u>HYPOTHESES ON THE COEFFICIENTS</u>: *Each triple* $(A^\lambda, B^\lambda, C^\lambda)$ *is graded according to* $\mathbb{R}^{d_1}\times\ldots\times\mathbb{R}^{d_q}$. *Moreover, let* $\eta\in\bigcap_{2\leq p<\infty}L^p(E,G)$, *and let* F^λ *be any one of the functions* $F^\lambda(y,\omega,t,z) = A^\lambda(y,\omega,t)$, *or* $B^{\lambda,\circ k}(y,\omega,t)$, *or* $C^\lambda(y,\omega,t,z)/\eta(z)$; *we suppose that:*

(i) F^λ *is of class* C^1 *in* y *on* $\mathbb{R}^d$, *for all* $\lambda\in\Lambda$, *and twice differentiable in* y *for* $\lambda=0$;

(ii) $|D^r_{y^r}F^\lambda(y,\omega,t,z)| \leq Z_t(\omega)(1+|y|^\theta)$ *for* $r=0$, $r=1$, *and all* $\lambda\in\Lambda$, *and also for* $r=2$ *if* $\lambda=0$;

(iii) $|D^r_{y^r}F^\lambda(y,\omega,t,z)-D^r_{y^r}F^0(y,\omega,t,z)| \leq Z_t(\omega)(1+|y|^\theta)|\lambda|$ *for* $r=0,1$;

(iv) $|\frac{\partial}{\partial y_j}F^{\lambda,i}(y,\omega,t,z)| \leq \zeta$ *if* $M_{r-1}<i,j\leq M_r$ *for some* $r\leq q$;

where $\zeta,\theta\geq 0$ *are constants and* Z *is a predictable process such that* $|||Z|||_p<\infty$ *for all* $p\in[1,\infty)$.

5-24 <u>THEOREM</u>: *Assume 5-23 and also that*

(i) $(H^\lambda)_{\lambda\in\Lambda}$ *is F-differentiable at 0, with derivative* ∂H;

(ii) $\{A^\lambda(Y^0_-)\}_{\lambda\in\Lambda}$, $\{B^\lambda(Y^0_-)\}_{\lambda\in\Lambda}$, $\{\frac{1}{\eta(z)} C^\lambda(Y^0_-,z)\}_{\lambda\in\Lambda}$ *are F-differentiable at* 0, *with derivatives* ∂A, ∂B, $\partial(C/\eta)$ (also denoted $\partial C/\eta$, η is the same as in 5-23, and Y^0 is the solution to 5-22 for $\lambda=0$, which exists by 5-10).

Then, the family $(Y^\lambda)_{\lambda\in\Lambda}$ *is F-differentiable at* 0, *and its derivative* ∂Y *is the unique solution to the linear equation obtained by formal differentiation of 5-22, namely:*

$$\begin{aligned}\partial Y &= \partial H + [\partial A + D_y A^0(Y^0_-)\partial Y_-]*t \\ &\quad + [\partial B + D_y B^0(Y^0_-)\partial Y_-]*W + [\partial C + D_y C^0(Y^0_-)\partial Y_-]*\tilde{\mu}.\end{aligned} \tag{5-25}$$

Proof. Each $(A^\lambda, B^\lambda, C^\lambda)$ satisfies 5-9, and $|H^\lambda|^*_T \in \cap_{p<\infty} L^p$, hence Y^λ exists and $|Y^\lambda|^*_T \in \cap_{p<\infty} L^p$. 5-25 is a linear equation, hence it has a unique solution ∂Y. Set $V^\lambda = Y^\lambda - Y^0 - \partial Y.\lambda$; the only remaining thing to prove is that $\|(V^\lambda)^*_T\|_{L^p} = o(|\lambda|)$ as $\lambda\to 0$, for all $p<\infty$. Now, it is clear that V^λ satisfies an equation 5-22, with initial condition $K^\lambda = \sum_{1\le i\le 3} K^{\lambda,i}$ and coefficients $(\overline{A}^\lambda, \overline{B}^\lambda, \overline{C}^\lambda)$ given by:

$\overline{A}^\lambda(y,\omega,t) = D_y A^0(Y^0_{t-}(\omega),\omega,t)y$, and similarly for $\overline{B}^\lambda$, $\overline{C}^\lambda$;

$$K^{\lambda,1} = H^\lambda - H^0 - \partial H.\lambda;$$

$$\begin{aligned}K^{\lambda,2} &= [A^\lambda(Y^0_-) - A^0(Y^0_-) - \partial A.\lambda]*t + [B^\lambda(Y^0_-) - B^0(Y^0_-) - \partial B.\lambda]*W \\ &\quad + [C^\lambda(Y^0_-) - C^0(Y^0_-) - \partial C.\lambda]*\tilde{\mu};\end{aligned}$$

$$\begin{aligned}K^{\lambda,3} &= [A^\lambda(Y^\lambda_-) - A^\lambda(Y^0_-) - D_y A^0(Y^0)(Y^\lambda_- - Y^0_-)]*t \\ &\quad + [B^\lambda(Y^\lambda_-) - B^\lambda(Y^0_-) - D_y B^0(Y^0_-)(Y^\lambda_- - Y^0_-)]*W \\ &\quad + [C^\lambda(Y^\lambda_-) - C^\lambda(Y^0_-) - D_y C^0(Y^0_-)(Y^\lambda_- - Y^0_-)]*\tilde{\mu}.\end{aligned}$$

$(\overline{A}^\lambda, \overline{B}^\lambda, \overline{C}^\lambda)$ obviously satisfy 5-9 with $Z_t = 0$ and $\hat{Z}_t = Z_t(1+|Y^0|^{*\theta}_{t-})$ and ζ as in 5-23 and $\theta=1$. Hence 5-11 gives $\|(V^\lambda)^*_T\|_{L^p} \leq c_p \|(K^\lambda)^*_T\|_{L^{\gamma_p}}$ for some c_p, γ_p independent of λ, and thus it suffices to prove that $\|(K^\lambda)^*_T\|_{L^p} = o(|\lambda|)$ for all $p<\infty$. That $\|(K^{\lambda,1})^*_T\|_{L^p} = o(|\lambda|)$ follows from (i); that $\|(K^{\lambda,2})^*_T\|_{L^p} = o(|\lambda|)$ follows from (ii) and an application of Lemma 5-1.

Consider now $\beta^\lambda = C^\lambda(Y^\lambda_-) - C^\lambda(Y^0_-) - D_y C^0(Y^0_-)(Y^\lambda_- - Y^0_-)$. Taylor's formula gives a function $\tilde{Y}^\lambda_t(\omega,z)$ such that $C^\lambda(Y^\lambda_{t-},\omega,t,z) = C^\lambda(Y^0_{t-},\omega,t,z) + D_y C^\lambda(\tilde{Y}^\lambda_t(\omega,z),\omega,t,z) \times (Y^\lambda_{t-}(\omega) - Y^0_{t-}(\omega))$, and which lies on the line segment joining $Y^\lambda_{t-}(\omega)$ and $Y^0_{t-}(\omega)$. Hence

$$\beta^\lambda = [D_y C^\lambda(\tilde{Y}^\lambda) - D_y C^0(\tilde{Y}^\lambda)](Y^\lambda_- - Y^0_-) + [D_y C^0(\tilde{Y}^\lambda) - D_y C^0(Y^0_-)](Y^\lambda_- - Y^0_-).$$

Therefore 5-23-(ii,iii) yields

$$\left|\frac{\beta^\lambda(t,z)}{\eta(z)}\right| \leq 2^{\theta-1} Z_t(1+|Y^0|^{*\theta}_{t-} + |Y^\lambda|^{*\theta}_{t-}) [|\lambda||Y^\lambda_{t-} - Y^0_{t-}| + |Y^\lambda_{t-} - Y^0_{t-}|^2],$$

and similar majorations for A^λ and B^λ instead of C^λ/η. Hence 5-1 yields

$$\|(K^{\lambda,3})^*_T\|_{L^p} \leq 2^{\theta-1} \delta_p |||Z|||_{3p} [1 + \|(Y^0)^{*\theta}_T\|_{L^{3p}} + \|(Y^\lambda)^{*\theta}_T\|_{L^{3p}}][|\lambda| \|(Y^\lambda - Y^0)^*_T\|_{L^{3p}} + \|(Y^\lambda - Y^0)^*_T\|^2_{L^{3p}}]. \tag{5-26}$$

Finally, 5-19 applied to Y^λ and Y^0 yields (recall that 5-23-(iii) holds for r=0):

$$\|(Y^\lambda - Y^0)^*_T\|_{L^q} \leq c'_q [\|(H^\lambda - H^0)^*_T\|_{L^{\gamma'_q}} + |\lambda| |||Z|||_{\gamma'_q}],$$

which is $O(|\lambda|)$ (apply (i)). Hence we deduce from 5-26 that $\|(K^{\lambda,3})^*_T\|_{L^p} = o(|\lambda|)$, and we are finished.

§5-c. PEANO'S APPROXIMATION

Now we wish to approximate the solution of 5-3, in which for simplicity we assume that $H \equiv y$, by the solutions to the following equations:

$$Y^n = y + A^n(Y^n_{\phi^n})*t + B^n(Y^n_{\phi^n})*W + C^n(Y^n_{\phi^n})*\tilde{\mu}, \quad (5\text{-}27)$$

where (A^n,B^n,C^n) are as in 5-4, and

$$\phi^n_t = \begin{cases} 0 & \text{if} \quad t=0 \\ kT2^{-n} & \text{if} \quad kT2^{-n}<t\leq(k+1)T2^{-n}. \end{cases} \quad (5\text{-}28)$$

To simplify our notation, we also write $(A^\infty,B^\infty,C^\infty)=(A,B,C)$ and $Y^\infty=Y$.

5-29 HYPOTHESES ON THE COEFFICIENTS: *Each triple* (A^n,B^n,C^n) *is graded according to* $\mathbb{R}^{d_1}\times\ldots\times\mathbb{R}^{d_q}$. *Moreover, let* $\eta\in\cap_{2\leq p<\infty}L^p(E,G)$, *and if* F^n *is any one of the functions* $F^n(y,\omega,t,z) = A^n(y,\omega,t)$, *or* $B^{n,\cdot k}(y,\omega,t)$, *or* $C^n(y,\omega,t,z)/\eta(z)$, *then*

(i) F^n *is differentiable in* y *on* $\mathbb{R}^d$;

(ii) $|F^n(0,\omega,t,z)| \leq Z^n_t(\omega)$ *for* $n\in\overline{\mathbb{N}}$.

(iii) $|D_yF^n(y,\omega,t,z)| \leq Z^n_t(\omega)(1+|y|^\theta)$ *for* $n\in\overline{\mathbb{N}}$;

(iv) $|F^n(y,\omega,t,z)-F^\infty(y,\omega,t,z)| \leq Z'^n_t(\omega)(1+|y|^\theta)$ *for* $n\in\mathbb{N}$;

(v) $|\frac{\partial}{\partial y_j}F^{n,i}(y,\omega,t,z)| \leq Z''^n_t(\omega)$ *if* $n\in\mathbb{N}$ *and* $M_{r-1}<i,j\leq M_r$ *for some* r;

(vi) $|\frac{\partial}{\partial y_j}F^{\infty,i}(y,\omega,t,z)| \leq\zeta$ *if* $M_{r-1}<i,j\leq M_r$ *for some* $r\leq q$,

where $\zeta,\theta\geq0$ *are constants and* Z^n, Z'^n, Z''^n *are predictable processes such that for all* $p<\infty$, $\sup_{n\in\overline{\mathbb{N}}}|||Z^n|||_p<\infty$, *and* $\beta_p < \infty$, *where*

$$\beta_p = \sup_{n\in\mathbb{N}} \sup_{t\leq T} \| E[(Z''^n_t)^p | \underline{\underline{F}}_{\phi^n_t}] \|_{L^\infty}. \qquad (5\text{-}30)$$

5-31 THEOREM: *Assume 5-29 and* $\|Z'^n\|_p \to 0$ *as* $n\uparrow\infty$ *for all* $p\in[1,\infty)$. *Then 5-3* (with $H\equiv y$) *and 5-27 have unique solutions, and*

$$\| (Y^n - Y)^*_T \|_{L^p} \to 0 \quad \text{as } n\uparrow\infty \text{ for all } p<\infty. \qquad (5\text{-}32)$$

Proof. a) (A,B,C) satisfy 5-9, so by Theorem 5-10, 5-3 has a unique solution Y and $Y^*_T \in \cap_{p<\infty} L^p$.

b) In order to prove that 5-27 has a solution, we begin with the case when the grading is trivial ($q=1$). Let $\psi^n_k = kT2^{-n}$. We construct Y^n by induction as follows: start with $Y^n_0 = y$. If $Y^n_{\psi^n_k}$ is known and belongs to $\cap_{p<\infty} L^p$, set for $\psi^n_k < t < \psi^n_{k+1}$:

$$Y^n_t = Y^n_{\psi^n_k} + \int_{\psi^n_k}^t A^n(Y^n_{\psi^n_k}, s)\,ds \qquad (5\text{-}33)$$

$$+ \int_{\psi^n_k}^t B^n(Y^n_{\psi^n_k}, s)\,dW_s + \int_{\psi^n_k}^t \int_E C^n(Y^n_{\psi^n_k}, s, z)\,\tilde{\mu}(ds,dz).$$

To see that 5-33 makes sense, we first remark that by 5-29-(ii,v):

$$|F^n(Y^n_{\psi^n_k}, t, z)| \leq Z^n_t + Z''^n_t |Y^n_{\psi^n_k}|,$$

(recall that $q=1$). If $H^n_s = [Z^n_s + Z''^n_s |Y^n_{\psi^n_k}|]\, 1_{(\psi^n_k, \psi^n_{k+1}]}(s)$, H^n is predictable, and

$$\int_{\psi^n_k}^t E(|H^n_s|^p)\,ds \leq 2^{p-1} \int_{\psi^n_k}^t E(|Z^n_s|^p + |Z''^n_s|^p\ |Y^n_{\psi^n_k}|^p)\,ds$$

$$\leq 2^{p-1} \int_{\psi^n_k}^t E(|Z^n_s|^p + |Y^n_{\psi^n_k}|^p\ E(|Z''^n_s|^p | \underline{\underline{F}}_{\psi^n_k}))\,ds \qquad (5\text{-}34)$$

$$\leq 2^{p-1}\int_{\psi_k^n}^t E(|Z_s^n|^p + \beta_p|Y_{\psi_k^n}^n|^p)ds < \infty$$

(because of 5-30 and $|||Z^n|||_p<\infty$). Hence, due to Lemma 5-1, 5-33 makes sense and

$$\sup_{\psi_k^n<t\leq\psi_{k+1}^n} |Y_t^n - Y_{\psi_k^n}^n| \in \cap_{p<\infty} L^p.$$

Then, an induction on k shows that Y^n is well-defined on $[0,T]$ and that $|Y^n|_T^*\in\cap_{p<\infty}L^p$.

c) In the general case, we can reproduce the proof of part (b) of Theorem 5-10, provided we replace everywhere Y_-, $\Pi_\ell Y_-$, $Q_{\ell+1}X_-$ by $Y^n_{\phi^n}$, $\Pi_\ell Y^n_{\phi^n}$, $Q_{\ell+1}X^n_{\phi^n}$. We obtain that 5-27 has a solution which satisfies $|Y^n|_T^*\in\cap_{p<\infty}L^p$. Moreover, any solution of 5-27 must satisfy 5-33 for $\psi_k^n<t\leq\psi_{k+1}^n$, so another induction on k shows that the solution to 5-27 is indeed unique.

We can further reproduce the proof of part (c) of Theorem 5-10 as well: replace " $|A(Y_-)|\leq Z+\zeta|Y_-|$ " by " $|A^n(Y^n_{\phi^n})|\leq Z^n+Z''^n|Y^n_{\phi^n}|$ "; the same trick as in 5-34 allows to obtain 5-14, with β_p instead of ζ^p. The rest of the proof goes exactly the same way, with $\hat{Z}=Z^n$. Therefore we obtain the estimate 5-11, and since $\sup_n|||Z^n|||_p<\infty$ for all $p<\infty$, we get

$$\sup_{n\in\mathbb{N}} \|(Y^n)_T^*\|_{L^p} < \infty \qquad \text{for all } p<\infty. \tag{5-35}$$

d) Finally, $\overline{Y}^n=Y^n-Y$ is a solution to an equation 5-3, with

$$\overline{H}^n = [A^n(Y^n_{\phi^n})-A(Y^n_{\phi^n})]*t + [B^n(Y^n_{\phi^n})-B(Y^n_{\phi^n})]*W + [C^n(Y^n_{\phi^n})-C(Y^n_{\phi^n})]*\tilde{\mu}$$

$$\overline{A}^n(y,\omega,t) = A(Y^n_{\phi^n_t}(\omega),\omega,t) - A(Y^n_{t-}(\omega)-y,\omega,t)$$

and the same for $\overline{B}^n$ and $\overline{C}^n$. These coefficients satisfy

5-9-(i) with $\bar{Z}^n_t = Z^\infty_t(1+|Y^n|^{*\theta}_{t-})|Y^n_{t-} - Y^n_{\phi^n_t}|$, and 5-9-(ii) with $\hat{Z}^n_t = Z_t$, and 5-9-(iii) with ζ. Hence 5-11 yields

$$\|(Y^n-Y)^*_T\|_{L^p} \le c_p(\|(\bar{H}^n)^*_T\|_{L^{\gamma_p}} + |||\bar{Z}^n|||_{\gamma_p})$$

for some c_p, γ_p. From $|||Z'^n|||_p \to 0$, 5-29-(iv) and 5-35, plus the usual application of Lemma 5-1, we deduce that $\|(\bar{H}^n)^*_T\|_{L^p} \to 0$ for all $p<\infty$. So it remains to prove $|||\bar{Z}^n|||_p \to 0$ for all $p<\infty$. From $|||Z^\infty|||_p<\infty$ for all $p<\infty$ and from 5-35, we deduce that it suffices to prove that $|||Y^n_- - Y^n_{\phi^n}|||_p \to 0$.

Now, we have seen that 5-11 holds for Y^n; it also holds for $Y^n_t - Y^n_{\phi^n_t}$, which yields

$$\|Y^n_t - Y^n_{\phi^n_t}\|_{L^p} \le c_p\{\int_{\phi^n_t}^t E(|Z^n_s|^{\gamma_p})ds\}^{1/\gamma_p}.$$

Thus $\|Y^n_t - Y^n_{\phi^n_t}\|_{L^p}$ clearly goes to 0 as $n\uparrow\infty$ (because $\phi^n_t\uparrow t$). We deduce that

$$\int_0^T E(|Y^n_{t-} - Y^n_{\phi^n_t}|^p)dt = \int_0^T E(|Y^n_t - Y^n_{\phi^n_t}|^p)dt \to 0$$

and this finishes the proof.

CHAPTER III

BISMUT'S APPROACH

Section 6: CALCULUS OF VARIATIONS

§6-a. THE GENERAL SETTING

In this section we fix $m\in\mathbb{N}^*$ and E, an open subset of $\mathbb{R}^\beta$ ($\beta\in\mathbb{N}^*$), with its Borel σ-field $\underline{\underline{E}}$. $G(dz)=dz$ denotes Lebesgue measure on $(E,\underline{\underline{E}})$.

Ω is the *canonical space* of all points $\omega=(w,\eta)$, where: - w is a continuous function:$[0,T]\to\mathbb{R}^m$, $w(0)=0$;

- η is an integer-valued measure on $[0,T]\times E$,

and $W_t(w,\eta)=w(t)$, $\mu((w,\eta);.)=\eta(.)$. Furthermore, $(\underline{\underline{F}}_t)_{0\leq t\leq T}$ is the smallest right-continuous filtration on Ω for which W and μ are optional, and $\underline{\underline{F}}=\underline{\underline{F}}_T$. We endow $(\Omega,\underline{\underline{F}})$ with the unique probability measure P such that

$$\left.\begin{array}{l} W \text{ is a standard m-dimensional Brownian motion;} \\ \mu \text{ is a Poisson measure with intensity } \nu(dt,dx)=dt\times G(dz) \\ W \text{ and } \mu \text{ are independent,} \end{array}\right\} \qquad (6\text{-}1)$$

and we write $\tilde{\mu}=\mu-\nu$ for the compensated Poisson measure.

We are given a point x_0 in $\mathbb{R}^d$, and a family of coefficients:

$a=(a^i)_{1\leq i\leq d}$: $\mathbb{R}^d\to\mathbb{R}^d$

$b=(b^{ij})_{1\leq i\leq d,1\leq j\leq m}$: $\mathbb{R}^d\to\mathbb{R}^d\otimes\mathbb{R}^m$

$c=(c^i)_{1\leq i\leq d}$: $\mathbb{R}^d\times E\to\mathbb{R}^d$, and we consider the following equation:

$$X = x_0 + a(X_-)*t + b(X_-)*W + c(X_-)*\tilde{\mu}. \tag{6-2}$$

We will obviously need some regularity assumptions on these coefficients for the purpose of the "iteration" outlined in Section 4; these conditions will be slightly less stringent than (A-r), and we set ($r\in\mathbb{N}^*$):

6-3 ASSUMPTION (A'-r): *The coefficients* (a,b,c) *are graded according to* $\mathbb{R}^{d_1}\times\ldots\times\mathbb{R}^{d_q}$ (see 5-5). *They are* r *times differentiable, and there are two constants* $\zeta,\theta>0$ *and a function* $\eta\in\cap_{2\leq p<\infty}L^p(E,G)$ *such that*

(i) $|D^n_{x^n}a(x)|$, $|D^n_{x^n}b(x)|$, $|\frac{1}{\eta(z)}D^n_{x^n}c(x,z)|$
$\leq \zeta(1+|x|^\theta)$ for $0\leq n\leq r$;

(ii) $|D^{n+k}_{x^n z^k}\, c(x,z)| \leq \zeta(1+|x|^\theta)$, $1\leq n+k\leq r$ *and* $k\geq 1$;

(iii) $|\frac{\partial^n}{\partial x_{i_1}\ldots\partial x_{i_n}}\alpha^i(x,z)|\leq\zeta$ *for* $1\leq n\leq r$, *if* α *is one of the functions* a(x), *or* $b^{\cdot k}(x)$, *or* $\frac{c(x,z)}{\eta(z)}$, *and if* $M_{s-1}<i,i_1,\ldots,i_n\leq M_s$ *for some* $s\leq q$ (recall that $M_0=0$, and $M_s=d_1+..+d_s$);

(iv) $|\frac{\partial^n}{\partial x_{i_1}\ldots\partial x_{i_n}}D^k_{z^k}\, c^i(x,z)|\leq\zeta$ *for* $1\leq n+k\leq r$ *and* $k\geq 1$, *and if* $M_{s-1}<i,i_1,\ldots,i_n\leq M_s$ *for some* $s\leq q$.

We always assume (A'-1) at least. Thus 6-2 has a unique solution X and $X^*_T\in\cap_{p<\infty}L^p$ in virtue of 5-10.

Now we briefly describe *Bismut's way of obtaining 4-4*. The idea consists in constructing, for each λ belonging to a neighbourhood Λ of 0 in $\mathbb{R}^d$:

1) a probability measure P^λ (with $P^0=P$) equivalent to P. Denote by $G^\lambda=dP^\lambda/dP$ the Radon-Nikodym derivative,

2) a "shift" θ^λ on Ω (with θ^0= identity), such that $P^\lambda \circ (\theta^\lambda)^{-1} = P$.

Then, if $Z \in L^1(P)$,

$$E[(Z\circ\theta^\lambda)G^\lambda] = E^\lambda(Z\circ\theta^\lambda) = E(Z). \tag{6-4}$$

Suppose that $(G^\lambda)_{\lambda\in\Lambda}$ is F-differentiable at 0 (see 5-21) and call H the set of all r.v. Z for which $(Z\circ\theta^\lambda)_{\lambda\in\Lambda}$ is F-differentiable at 0. Let $\Phi=(\Phi^1,..,\Phi^d)$ be the variable of interest, and suppose also that $\Phi^i \in H$. Then indeed the following is an *integration-by-parts setting for* Φ (see 4-4):

$$\left.\begin{aligned} &\sigma_\Phi = \partial\Phi \quad (\text{i.e. } \sigma_\Phi^{ij} = \frac{\partial\Phi^i\circ\theta^\lambda}{\partial\lambda_j}\Big|_{\lambda=0}), \\ &\gamma_\Phi = -\partial G, \quad H_\Phi = H \\ &\delta_\Phi(\Psi) = -\partial\Psi \quad \text{for } \Psi\in H \end{aligned}\right\} \tag{6-5}$$

(4-5 is trivial, and 4-6 follows from differentiating 6-4 at $\lambda=0$, with $Z=\Psi f(\Phi)$).

Of course, there are many choices for θ^λ and P^λ (e.g. $\theta^\lambda\equiv$identity and $P^\lambda=P$). But the shifts θ^λ should in fact be chosen so as to make σ_Φ invertible for $\Phi=X_t$: so one sees on 6-5 why the construction 4-4 in this method depends upon the solution X itself.

§6-b. THE GIRSANOV TRANSFORM

First, we construct the shifts θ^λ on Ω, indexed by the elements of a neighbourhood Λ of 0 in $\mathbb{R}^d$. We shall denote by $W^\lambda=W\circ\theta^\lambda$ and $\mu^\lambda=\mu\circ\theta^\lambda$ the "perturbed" driving terms.

The idea is that the law of (W^λ,μ^λ) be equivalent with the law of (W,μ). It is well known that this can be achieved for W^λ by setting:

6-6 $W_t^\lambda = W_t + \int_0^t u_s \circ \lambda \, ds$, where $u = (u^{ij})_{1 \leq i \leq m, 1 \leq j \leq d}$ is an $\mathbb{R}^m \otimes \mathbb{R}^d$-valued predictable process, and $u.\lambda = (\sum_j u^{ij} \lambda^j)_{1 \leq i \leq m}$.

Then $W^0 = W$ and the perturbation is linear in λ. As we need first derivatives only at $\lambda = 0$, this is good enough. As for μ^λ, infinitely many jumps have to be perturbed simultaneously (for reasons apparent later); this essentially requires that the jump times of μ be left unchanged and that only the *jump sizes* be modified. We can do as follows:

6-7 $\mu^\lambda = \gamma^\lambda(\mu)$ is the image of $\mu(\omega;.)$ under the map $(t,z) \rightsquigarrow (t, \gamma^\lambda(\omega,t,z))$, where $\gamma^\lambda(\omega,t,z) = z + v(\omega,t,z).\lambda$ (i.e., $\gamma^\lambda(\omega,t,z)^i = z^i + \sum_{j \leq d} v^{ij}(\omega,t,z)\lambda^j$ for $i \leq \beta$). In this, $v = (v^{ij})_{1 \leq i \leq \beta, 1 \leq j \leq d}$ is an $\mathbb{R}^\beta \otimes \mathbb{R}^d$-valued function on $\Omega \times [0,T] \times E$ which is $\underline{\underline{P}} \otimes \underline{\underline{E}}$-measurable. (Recall that $\underline{\underline{P}}$ is the predictable σ-field).

Then, Ω being the canonical space, the map $\theta^\lambda : \Omega \to \Omega$ is entirely defined by

$$W \circ \theta^\lambda = W^\lambda, \quad \mu \circ \theta^\lambda = \mu^\lambda. \tag{6-8}$$

For technical reasons, we will need boundedness assumptions on u and v. First, we introduce an auxiliary function $\rho : E \to [0,\infty)$, which meets the following (the same as 2-23):

6-9 *Conditions on* ρ: (i) ρ is of class C_b^∞, with $|D_{z^r}^r \rho| \in L^1(E,G)$ for all $r \in \mathbb{N}$.

(ii) $\rho(z) \to 0$ as z goes to the boundary of E.

6-10 *Conditions on* u *and* v: (i) u is bounded;

(ii) $z \rightsquigarrow v(\omega,t,z)$ is differentiable, and $|v(\omega,t,z)| \leq \rho(z)$ and $|D_z v(\omega,t,z)| \leq \rho(z)$.

Observe that 6-9 implies that $\rho(z) \leq a\, d(z,\partial E)$, where $d(z,\partial E)$ denotes the distance of z to the boundary of E, and $a := \sup|D_z \rho(z)|$. Then we easily deduce from 6-10-(ii) that, if Λ is small enough, for all ω,t,λ,

$$z \rightsquigarrow \gamma^\lambda(\omega,t,z) \text{ is a bijection from E onto E.} \tag{6-11}$$

The first order of business is to find a Radon-Nikodym derivative G^λ such that if $P^\lambda = G^\lambda.P$, then $P^\lambda \circ (\theta^\lambda)^{-1} = P$. For this, we first enter the Jacobian of the transformation 6-11, to wit

$$Y^\lambda = \det(I + \textstyle\sum_{i\leq d} D_z v^{.i}.\lambda^i) \tag{6-12}$$

(I is the identity matrix with the correct size, namely $\beta\times\beta$ here). Y is a polynomial in λ, of degree β, whose constant term is 1, and whose other coefficients are themselves polynomials in $D_z v$. In particular, the coefficient of λ^i in the degree 1 term is

$$\operatorname{div} v^{.i} := \textstyle\sum_{j\leq\beta} \frac{\partial}{\partial z_j} v^{ji}. \tag{6-13}$$

In view of 6-9 and 6-10, the previous discussion allows to shrink further Λ so that for all $\lambda\in\Lambda$,

$$|Y^\lambda - 1| \leq \tfrac{1}{2}\wedge(|\lambda|k\rho), \quad |Y^\lambda - 1 - \operatorname{div} v.\lambda| \leq |\lambda|^2 k\rho \tag{6-14}$$

(where $\operatorname{div} v.\lambda = \sum_{j\leq\beta,\, i\leq d} \frac{\partial}{\partial z_j} v^{ji}\lambda^i$), for some constant k.

As u is bounded and $\rho\in L^1(E,G)$ and 6-14 holds, the following defines a (real-valued) martingale:

$$M^\lambda = -(u.\lambda)*W + (Y^\lambda - 1)*\tilde{\mu}.$$

Its Doléans-Dade exponential $G^\lambda = \mathcal{E}(M^\lambda)$ is the solution to the linear equation

$$G^\lambda = 1 - [G^\lambda_-(u.\lambda)]*W + [G^\lambda_-(Y^\lambda-1)]*\tilde{\mu}. \tag{6-15}$$

6-16 THEOREM: $P^\lambda(d\omega) = G^\lambda_T(\omega).P(d\omega)$ *is a probability measure on* $(\Omega,\underline{\underline{F}})$ *and satisfies* $P = P^\lambda \circ (\theta^\lambda)^{-1}$.

Proof. a) 6-15 is a 1-dimensional equation of type 5-22 with $H^\lambda = 1$, $A^\lambda = 0$, $B^\lambda(y,\omega,t) = -yu_t(\omega).\lambda$, $C^\lambda(y,\omega,t,z) = y(Y^\lambda(\omega,t,z)-1)$. Since u is bounded and 6-14 holds, these coefficients trivially fulfill 5-23 with $\eta=\rho$ and 5-24-(i,ii), with $\partial A=0$, $\partial B=-u$, $\partial C=\text{div}\, v$ (due to 6-14). Theorem 5-24 yields that (G^λ) is F-differentiable, and in particular $|G^\lambda|^*_T \in \cap_{p<\infty} L^p$: hence the local martingale G^λ is a martingale, and $E(G^\lambda_T)=1$.

Moreover, 6-14 yields that $|\Delta M^\lambda| \leq 1/2$, thus $G^\lambda_T > 0$ a.s.; therefore $P^\lambda = G^\lambda_T.P$ is indeed a probability measure, equivalent to P.

b) For the second claim, we apply Girsanov's Theorem for continuous martingales [13,(7-25a)] and for random measures [13,(7-31)] as follows:

(i) Under P^λ, W^λ is a continuous local martingale starting at 0 and having the same quadratic variation as W; hence W^λ is a standard m-dimensional Brownian motion under P^λ.

(ii) Under P^λ, the measure μ has $Y^\lambda.\nu$ for its dual predictable projection (or, compensator). Since γ^λ is predictable by construction, $\mu^\lambda = \gamma^\lambda(\mu)$ has the image $\gamma^\lambda(Y^\lambda.\nu)$ of $Y^\lambda.\nu$ for its dual predictable projection under P^λ. Now, 6-11, 6-12 and the classical change of variables formula for Lebesgue measure on E give for every Borel subset A of $[0,T]\times E$:

$$[\gamma^\lambda(Y^\lambda.\nu)](A) = \int_0^T ds \int_E dz\ 1_A(s,z+v(s,z).\lambda)Y^\lambda(s,z)$$
$$= \int_0^T ds \int_E dz\ 1_A(s,z) = \nu(A).$$

Hence $\gamma^\lambda(Y^\lambda.\nu)=\nu$ is the dual predictable projection of μ^λ under P^λ, that is, μ^λ is a Poisson measure under P^λ, with intensity ν.

Thus (W^λ,μ^λ) has the same distribution under P^λ as (W,μ) under P. Since P is the unique measure meeting 6-1 and since $(W^\lambda,\mu^\lambda)=(W,\mu)\circ\theta^\lambda$, it follows that $P^\lambda\circ(\theta^\lambda)^{-1}=P$.

From part (a) of the proof above, we also deduce:

6-17 COROLLARY: *The family* (G^λ) *is F-differentiable at* 0, *and its derivative* ∂G *is given by formal differentiation of 6-15:*

$$\partial G = -u*W + (\operatorname{div} v)*\tilde{\mu} \tag{6-18}$$

(i.e., $\partial G^i = -\sum_{j\leq m} u^{ji}*W^j + (\sum_{j\leq\beta}\frac{\partial}{\partial z_j}v^{ji})*\tilde{\mu}$).

§6-c. PERTURBATION OF THE STOCHASTIC DIFFERENTIAL EQUATION

So far we have obtained 6-4, and we need now to prove that the family $\{X_t\circ\theta^\lambda\}$ is F-differentiable. Formally, $Y^\lambda=X\circ\theta^\lambda$ should be solution to

$$Y^\lambda = x_0 + a(Y^\lambda_-)*t + b(Y^\lambda_-)*W^\lambda + c(Y^\lambda_-)*(\mu^\lambda-\nu). \tag{6-19}$$

Of course, ν is not the compensator of μ^λ under P, so this equation is not of the form studied in Section 5. Nevertheless, we have:

6-20 LEMMA: *Under (A'-1),* $Y^\lambda = X \circ \theta^\lambda$ *satisfies*

$$Y^\lambda = x_0 + a(Y^\lambda_-)*t + b(Y^\lambda_-)*W + b(Y^\lambda_-)u\lambda*t \\ + c(Y^\lambda_-,\gamma^\lambda(z))*\tilde{\mu} + [c(Y^\lambda_-,\gamma^\lambda(z)) - c(Y^\lambda_-,z)]*\nu. \tag{6-21}$$

Proof. We consider successively $U=a(X_-)*t$, $U=b(X_-)*W$ and $U=c(X_-)*\tilde{\mu}$.

a) If $U=a(X_-)*t$, then $U\circ\theta^\lambda = a(Y^\lambda_-)*t$ is trivial.

b) Let $U=b(X_-)*W$ and $U^\lambda = U\circ\theta^\lambda$. Since $P=P^\lambda\circ(\theta^\lambda)^{-1}$ and $\underline{\underline{F}}_t \subset (\theta^\lambda)^{-1}(\underline{\underline{F}}_t)$, classical results concerning mappings of processes (see e.g. [13,(10-38),(10-34),(10-40)]) imply that U^λ is a P^λ-local martingale of the form $b(X_-\circ\theta^\lambda)*(W\circ\theta^\lambda) = b(Y^\lambda_-)*W^\lambda$. As $P^\lambda \sim P$, the stochastic integrals are the same under P and P^λ, so $U^\lambda = b(Y^\lambda_-)*W + b(Y^\lambda_-)u.\lambda*t$.

c) Let $U=c(X_-)*\tilde{\mu}$ and $U^\lambda = U\circ\theta^\lambda$. Like in (b) we obtain that U^λ is a P^λ-purely discontinuous local martingale, and since $P^\lambda \sim P$, U^λ is a P-semimartingale with no continuous martingale part. Now, using 6-16 and the property $G^\lambda_T \in \cap_{p<\infty} L^p(P)$, we obtain for every r.v. Z:

$$\|Z\circ\theta^\lambda\|_{L^p(P)} = \|Z\|_{L^p(P^\lambda)} \leq \|Z\|_{L^{2p}(P)} \|G^\lambda_T\|_{L^2(P)} \tag{6-22}$$

Since $U^*_T \in \cap_{p<\infty} L^p(P)$ by 5-1, we deduce from 6-22 that $|U^\lambda|^*_T \in \cap_{p<\infty} L^p(P)$; thus the P-semimartingale U^λ is special (see [13] or [20]), and it is also quasi-left-continuous because it jumps only when μ jumps. Hence $U^\lambda = M+A$ where M is a P-purely discontinuous local martingale and A is continuous with finite variation. Identifying the jumps $\Delta M = \Delta U^\lambda$, we get that

$M = c(Y_-^\lambda, \gamma^\lambda(z)) * \tilde{\mu}$ (see [13]).

It remains to identify A. Let f_n be the restriction to E of the function: $z \leadsto 1 \wedge (n-|z|)^+$, and $c_n(x,z) = c(x,z) f_n(z)$, $U(n) = c_n(X_-) * \tilde{\mu}$, $U^\lambda(n) = U(n) \circ \theta^\lambda$. Exactly as above we have $U^\lambda(n) = c_n(Y_-^\lambda, \gamma^\lambda(z)) * \tilde{\mu} + A(n)$. But if η, ζ, θ are as in (A'-1), $|c_n(X_-, z)| \leq \zeta(1+|X_-|^\theta)\eta(z) f_n(z)$, and $\eta f_n \in L^1(E,G)$; thus $U(n)$ is also $c_n(X_-)*\mu - c_n(X_-)*\nu$, both integrals making sense; hence $U^\lambda(n) = c_n(Y_-^\lambda, \gamma^\lambda(z)) * \tilde{\mu} - c_n(Y_-^\lambda, z) * \nu$. We deduce that

$$A(n) = [c_n(Y_-^\lambda, \gamma^\lambda(z)) - c_n(Y_-^\lambda, z)] * \nu.$$

Now, f_n is Lipschitz with constant 1. Hence (A'-1)-(i,ii) and 6-10 yield

$$|c_n(Y_-^\lambda, \gamma^\lambda(z)) - c_n(Y_-^\lambda, z)| \leq \zeta |\lambda| (1+|Y_-^\lambda|^\theta)(1+\eta(z))\rho(z)$$

and the right-hand side above is G-integrable. Since $c_n \to c$ pointwise as $n \uparrow \infty$ we deduce from Lebesgue convergence theorem that for all $t \leq T$:

$$A(n)_t \to [c(Y_-^\lambda, \gamma^\lambda(z)) - c(Y_-^\lambda, z)] * \nu_t. \tag{6-23}$$

Moreover, Lebesgue convergence theorem for stochastic integrals (or Lemma 5-1) also yields that $|U(n)-U|_T^* \to 0$ in all $L^p(P)$ ($p<\infty$), as well as $|M(n)-M|_T^* \to 0$. From 6-22 we deduce that $|U^\lambda(n) - U^\lambda|_T^* \to 0$ in $L^p(P)$, hence $|A(n)-A|_T^* \to 0$ in $L^p(P)$. From 6-23 it follows that $U = c(Y_-^\lambda, \gamma^\lambda(z)) * \tilde{\mu} + [c(Y_-^\lambda, \gamma^\lambda(z)) - c(Y_-^\lambda, z)] * \nu$.

Putting together the results of (a), (b), (c) yields 6-21.

6-24 <u>THEOREM</u>: *Under (A'-2) the family* $\{X \circ \theta^\lambda\}_{\lambda \in \Lambda}$ *is F-differentiable at* 0, *and its derivative* $DX := \frac{\partial}{\partial \lambda}(X \circ \theta^\lambda)_{|\lambda=0}$

is the solution to the linear equation obtained by formal differentiation of 6-21:

$$DX = D_x a(X_-)DX_- * t + D_x b(X_-)DX_- * W$$
$$+ D_x c(X_-)DX_- * \tilde{\mu} + b(X_-)u * t + D_z c(X_-) * \mu. \tag{6-25}$$

<u>Proof</u>. Due to 6-20, $Y^\lambda = X \circ \theta^\lambda$ is the solution to an equation 5-22, with $H^\lambda = x_0$ and the coefficients

$$\left.\begin{aligned} A^\lambda(y,\omega,t) &= a(y) + b(y)u_t(\omega)\lambda \\ &\quad + \int_E [c(y,z+v(\omega,t,z)\lambda) - c(y,z)]\,dz \\ B^\lambda(y,\omega,t) &= b(y) \\ C^\lambda(y,\omega,t,z) &= c(y,z+v(\omega,t,z)\lambda). \end{aligned}\right\} \tag{6-26}$$

We wish to show that the assumptions of Theorem 5-24 are met. This is trivial for 5-24-(i) (with $H=0$), and $(A^\lambda, B^\lambda, C^\lambda)$ is obviously graded according to the grading $\mathbb{R}^d = \mathbb{R}^{d_1} \times \ldots \times \mathbb{R}^{d_q}$. For the other conditions, it suffices to prove that they are met, separately for each of the following terms: $A_1^\lambda = a$, $A_2^\lambda = bu\lambda$, $A_3^\lambda(y,\omega,t) = \int_E \tilde{A}_3^\lambda(y,\omega,t,z)dz$, where $\tilde{A}_3^\lambda(y,\omega,t,z) = c(y,z+v(\omega,t,z)\lambda) - c(y,z)$, and B^λ, and C^λ.

Due to (A'-2) and 6-10, A_1^λ and A_2^λ and B^λ clearly satisfy 5-23 (with $Z_t(\omega) = \zeta + \zeta \sup_z \rho(z)$). The families $\{A_1^\lambda(X_-)\}$ and $\{B^\lambda(X_-)\}$ are obviously F-differentiable at 0, with derivatives $\partial A_1 = 0$, $\partial B = 0$. Since $X_T^* \in \cap_{p<\infty} L^p$, it is also easy to check that $\{A_2^\lambda(X_-)\}$ is F-differentiable at 0, with derivative $\partial A_2 = b(X_-)u$.

Consider now C^λ. We use the notation of (A'-2), and $k \geq 1$ satisfying $|\lambda| \leq k$ for all $\lambda \in \Lambda$. C^λ obviously satisfies (i), and (ii) for $\lambda = 0$, of 5-23, by (A'-2). Due to 6-10 and (A'-2) (first (ii), then (i), then (iii) and (iv)), we have:

$$\left.\begin{array}{l} |D^r_{y^r} C^\lambda(y,z) - D^r_{y^r} C^0(y,z)| \\ \qquad \leq \zeta|\lambda|(1+|y|^\theta)\rho(z) \quad \text{if } r=0,1; \\ |D^r_{y^r} C^\lambda(y,z)| \leq \zeta k(1+|y|^\theta)[\rho(z)+\eta(z)] \\ \qquad \text{if } r=0,1 \\ |\frac{\partial}{\partial y_j} C^{\lambda,i}(y,z)| \leq \zeta k[\rho(z)+\eta(z)] \\ \qquad \text{if } M_{s-1}<i,j\leq M_s \quad \text{for some} \quad s\leq q. \end{array}\right\} \quad (6\text{-}27)$$

Hence (C^λ) satisfies 5-23 with $\eta'=\rho+\eta$ (which belongs to $\cap_{2\leq p<\infty} L^p(E,G)$). Next we prove that the family $\{C^\lambda(X_-,z)/\eta'(z)\}$ is F-differentiable at 0, with derivative $\partial C/\eta'=D_z c(X_-)v/\eta'$. 6-27 yields $|C^\lambda(X_-)/\eta'| \leq k\zeta(1+|X_-|^\theta)$, while $|v|\leq\rho$ and (A'-2) yield $|\partial C/\eta'| \leq \zeta(1+|X_-|^\theta)\sup_z \rho(z)$, whose supremum in time belong to $\cap_{p<\infty} L^p$. We also have (use Taylor's formula and (ii) of (A'-2)):

$$\frac{1}{\eta'}|C^\lambda(X_-)-C^0(X_-)-\partial C.\lambda| \leq \frac{\zeta}{2}(1+|X_-|^\theta)|\lambda|^2 \sup_z \rho(z),$$

and we easily deduce the desired result.

It remains to consider A^λ_3. 5-23 is satisfied for $\lambda=0$, because $A^0_3=0$. Using the equality $\tilde{A}^\lambda_3=C^\lambda-C^0$ and 6-27 and the definition of A^λ_3, we deduce that A^λ_3 is indeed well-defined, is once differentiable, and that

$$|D^r_{y^r} A^\lambda_3(y,\omega,t)| \leq |\lambda|\zeta(1+|y|^\theta)G(\rho) \quad \text{for } r=0,1$$

($G(\rho)$ denotes the integral of ρ with respect to G). Hence 5-23-(i,ii,iii) holds with $Z_t(\omega)=\zeta k G(\rho)$. Using (iv) of (A'-2) and Taylor's formula, we see that $|\frac{\partial}{\partial y_j}\tilde{A}^{\lambda,i}_3(y,\omega,t,z)| \leq \zeta|\lambda|\rho(z)$ if $M_{s-1}<i,j\leq M_s$ for some $s\leq q$; hence $|\frac{\partial}{\partial y_j} A^{\lambda,i}_3(y,\omega,t)| \leq \zeta k G(\rho)$ and 5-23-(iv) is met.

Finally we prove that the family $\{A_3^\lambda(X_-)\}$ is F-differentiable at 0, with derivative $\partial A_3 = \int_E D_z c(X_-,z)v(z)dz$. Since $|A_3^\lambda(X_-)| \leq k\zeta G(\rho)(1+|X_-|^\theta)$ we have $|A_3^\lambda(X_-)|_T^* \in \cap_{p<\infty} L^p$, and also

$$A_3^\lambda(X_-) - A_3^0(X_-) - \partial A_3 . \lambda$$
$$= \int_E [c(X_-,z+v(z)\lambda) - c(X_-,z) - D_z c(X_-,z)v(z)\lambda]\,dz.$$

Since $|v| \leq \rho$, (ii) of (A'-2) and Taylor's formula give

$$|A_3^\lambda(X_-) - A_3^0(X_-) - \partial A_3 . \lambda| \leq \frac{\zeta}{2}|\lambda|^2 \; G(\rho^2)(1+|X_-|^\theta)$$

and we easily deduce the claimed result.

Finally, in view of the form of ∂A_1, ∂A_2, ∂A_3, ∂B, ∂C, Theorem 5-24 implies that $\{Y^\lambda\}$ is F-differentiable at 0 and its derivative DX has

$$\begin{aligned} DX = {} & D_x a(X_-)DX_- * t + D_x b(X_-)DX_- * W \\ & + D_x c(X_-)DX_- * \tilde{\mu} + b(X_-)u * t + D_z c(X_-)v * \tilde{\mu} \\ & + [\int_E D_z c(X_-,z)v(z)dz] * t. \end{aligned} \tag{6-28}$$

Now, the last term is equal to $D_z c(X_-)v * \nu$, and $|D_z c(X_-,z)v(z)|$ is clearly ν-integrable. So the sum of the last two terms in 6-28 is $D_z c(X_-)v * \tilde{\mu}$, and 6-28 reduces to 6-25.

§6-d. EXPLICIT COMPUTATION OF DX.

In this subsection, we make explicit the dependence of X upon the initial condition, writing X^{x_0} for the solution to 6-2.

6-29 THEOREM: *Under (A'-2) the family* $\{X^x\}_{x \in \mathbb{R}^d}$ *is F-*

differentiable at each point $x \in \mathbb{R}^d$, *and its derivative at* x, *denoted by* ∇X^x, *is the unique solution to the linear equation obtained by formal differentiation of 6-2:*

$$\nabla X^x = I + D_x a(X_-^x)\nabla X_-^x * t + D_x b(X_-^x)\nabla X_-^x * W + D_x c(X_-^x)\nabla X_-^x * \tilde{\mu}. \tag{6-30}$$

Proof. Apply 5-24 with $H^x \equiv x$, and coefficients $A^x = a$, $B^x = b$, $C^c = c$.

As is well known, one can "explicitely" compute ∇X^x as follows. For each x, introduce the following $\mathbb{R}^d \otimes \mathbb{R}^d$ -valued process:

$$K^x = D_x a(X_-^x) * t + D_x b(X_-^x) * W + D_x c(X_-^x) * \tilde{\mu} \tag{6-31}$$

(due to (A'-2) and Lemma 5-1, this is well defined). Then 6-30 reads as

$$(\nabla X^x)^T = I + (\nabla X_-^x)^T * (K^x)^T \tag{6-32}$$

(or, in differential form, $d(\nabla X^x) = dK^x \, \nabla X_-^x)$. The solution ∇X^x to 5-32 is the transpose of the *Doléans-Dade exponential* of the transpose of K^x.

Now, we come back to 6-25, writing DX^x if the initial condition is $x_0 = x$. Set

$$H^x = b(X_-^x)u * t + D_z c(X_-^x) v * \mu \tag{6-33}$$

(we will see presently that u, v also depend on x). Then 6-25 gives

$$DX^x = H^x + [(DX^x)^T * (K^x)^T]^T \qquad (\text{or: } d(DX^x) = dH^x + dK^x \, DX_-^x). \tag{6-34}$$

This is a linear equation associated with the homogeneous equation 6-32. Its solution can be explicitely computed in terms of H^x and ∇X^x, by the usual method of variation of parameters. Set

$$\left.\begin{aligned} &T_0^x = 0,\ T_{n+1}^x = \inf(t>T_n^x:\ \det(I+\Delta K_t^x)=0) \\ &K^x(n)_t = K_t^x - K^x_{t\wedge T_{n-1}^x} \\ &\nabla X^x(n)^T = I + (\nabla X^x(n)_-)^T * (K^x(n))^T \end{aligned}\right\} \quad (6\text{-}35)$$

(So $K^x(1)=K^x$ and $\nabla X^x(1)=\nabla X^x$). Since K^x is càdlàg, $T_n^x\uparrow\infty$ as $n\uparrow\infty$, and we know [14] that

$$\left.\begin{aligned} &\nabla X^x(n)_t = I \quad \text{if } t\le T_{n-1}^x, \\ &\nabla X^x(n)_t \text{ is invertible if } t<T_n^x \\ &\nabla X^x(n)_t \text{ is not invertible if } t\ge T_n^x, \\ &\nabla X^x(n)_{t-} \text{ is invertible if } t\le T_n^x. \end{aligned}\right\} \quad (6\text{-}36)$$

Then it is shown in [14] that, since H^x has finite variation,

$$\left.\begin{aligned} &DX_t^x = \nabla X^x(n)_t\ \{DX^x_{T_{n-1}^x} \\ &+ \int_{T_{n-1}^x}^t \nabla X^x(n)_{s-}^{-1}(I+\Delta K_s^x)^{-1}\, dH_s^x\} \quad \text{if } T_{n-1}^x\le t<T_n^x \\ &DX^x_{T_n^x} = \Delta H^x_{T_n^x} + (I+\Delta K^x_{T_n^x})DX^x_{(T_n^x)-} \qquad (\text{and } DX_0^x=0). \end{aligned}\right\} (6\text{-}37)$$

§6-e. HIGHER DERIVATIVES

We see on formula 6-25 that if we want to further differentiate $DX^x\circ\theta^\lambda$ in λ, we need to differentiate $u\circ\theta^\lambda$ and $v\circ\theta^\lambda$, so 6-10 is not sufficient. On the other hand, as already noticed before, the shifts θ^λ should depend on the solution X^x itself, and in particular on the starting point x: in other words, u and v might depend on x, and we write them u^x and v^x.

Instead of starting with arbitrary (differentiable enough) u^x, v^x, we restrict in fact to the following form, which is sufficient for our purposes:

6-38 *Conditions on* u^x *and* v^x: We have

$$u_t^x(\omega) = f(X_{t-}^x(\omega), \nabla X_{t-}^x(\omega)),$$

$$v^x(\omega,t,z) = g(X_{t-}^x(\omega), \nabla X_{t-}^x(\omega), z),$$

where (i) $f: \mathbb{R}^d \times (\mathbb{R}^d \otimes \mathbb{R}^d) \to \mathbb{R}^m \otimes \mathbb{R}^d$ is of class C_b^∞;

(ii) $g(x,y,z) = g_o(x,y,z)\rho(z)$, where ρ satisfies 6-9 and $g_o : \mathbb{R}^d \times (\mathbb{R}^d \otimes \mathbb{R}^d) \times E \to \mathbb{R}^\beta \otimes \mathbb{R}^d$ is of class C_b^∞.

Upon multiplying g by a constant, we can and will also assume that 6-10 is met. 6-38 and 6-9 clearly imply for all $n,k,\ell \in \mathbb{N}$:

$$\left.\begin{array}{l} z \rightsquigarrow \sup_{x,y} |D^{n+k+\ell}_{x^n y^k z^\ell} g(x,y,z)| \quad \text{belongs} \\ \text{to} \quad \bigcap_{1 \leq p < \infty} L^p(E,G). \end{array}\right\} \tag{6-39}$$

Then 6-15 defines $G^{x,\lambda}$ for each $x \in \mathbb{R}^d$; note that the shifts $(\theta^{x,\lambda})$ also depend on x and for each x the family $\{G^{x,\lambda}\}_{\lambda \in \Lambda}$ is F-differentiable at 0, with derivative

$$\partial G^x = -f(X_-^x, \nabla X_-^x) * W + \operatorname{div}_z g(X_-^x, \nabla X_-^x) * \tilde{\mu}, \tag{6-40}$$

while 6-33 reads as

$$H^x = b(X_-^x) f(X_-^x, \nabla X_-^x) * t + D_z c(X_-^x) g(X_-^x, \nabla X_-^x) * \mu. \tag{6-41}$$

As seen in the previous subsection, we are led to consider processes Y^x indexed by $x \in \mathbb{R}^d$. We have two kinds of derivatives:

6-42 The family $\{Y^x\}_{x\in\mathbb{R}^d}$ may be F-differentiable at each point x; the derivative is then denoted by ∇Y^x. If one can iterate the procedure r times, one says that Y^x is r *times F-differentiable* (in x) and we denote by $\nabla^k Y^x$ the successive derivatives (with $\nabla^0 Y^x = Y^x$).

6-43 The family $\{Y^x \circ \theta^{x,\lambda}\}_{\lambda\in\Lambda}$ may be F-differentiable at 0, for all x; the derivative is then denoted by DY^x. If we can iterate the procedure r times we say that Y^x is *F-(r)-differentiable* (in λ) and we denote by $D^k Y^x$ the successive derivatives (with $D^0 Y^x = Y^x$); for instance, $D^2 Y^x$ is the derivative of $\{(DY^x)\circ\theta^{x,\lambda}\}_{\lambda\in\Lambda}$.

Our aim in this subsection is to prove:

6-44 THEOREM: *Assume (A'-r) for some* $r\geq 3$, *and 6-38. Then a)* $\{X^x\}$ *is* (r-1) *times F-differentiable in* x, *and* $\{\partial G^x\}$ *is* (r-2) *times F-differentiable in* x.

b) If $0\leq k\leq r-2$, $\{\nabla^k X^x\}$ *is F-*(r-k-1)*-differentiable* (in λ: see 6-43)*;if* $0\leq k\leq r-3$, $\{\nabla^k \partial G^x\}$ *is F-*(r-k-2)*-differentiable* (in λ).

Furthermore, $\nabla^k D^n X^x = D^n \nabla^k X^x$ *for* $n+k\leq r-1$ *and* $\nabla^k D^n \partial G^x = D^n \nabla^k \partial G^x$ *for* $n+k\leq r-2$.

Proof. Set $\overline{X}^x = (X^x, \nabla X^x, DX^x, \partial G^x)$, which takes its values in $F = \mathbb{R}^d \times (\mathbb{R}^d \otimes \mathbb{R}^d) \times (\mathbb{R}^d \otimes \mathbb{R}^d) \times \mathbb{R}^d$. A point $\overline{x}$ of F is denoted by $\overline{x} = (x, y, w, \gamma)$ according to the above decomposition of F into four factors. If we put together 6-2, 6-25 (or, rather, 6-28), 6-30 and 6-40, we see that $\overline{X}^x$ is solution to an equation 6-2, with dimension

$\overline{d}=2d+2d^2$, with initial condition $\overline{x}=(x,I,0,0)$, and with the following coefficients: $\overline{a},\overline{b},\overline{c}$:

- if $i\leq d$: $\overline{a}^i(x,y,w,\gamma)=a^i(x)$, $\overline{b}^{in}(x,y,w,\gamma)=b^{in}(x)$,
  $\overline{c}^i((x,y,w,\gamma),z)=c^i(x,z)$.

- if $d<i\leq d+d^2$ and i corresponds to the $(j,k)^{th}$ coordinate in $\mathbb{R}^d\otimes\mathbb{R}^d$:
  $$\overline{a}^i(x,y,w,\gamma) = \sum_{\ell\leq d} \frac{\partial}{\partial x_\ell}a^j(x)y^{\ell k},$$
  $$\overline{b}^{in}(x,y,w,\gamma) = \sum_{\ell\leq d} \frac{\partial}{\partial x_\ell}b^{jn}(x)y^{\ell k},$$
  $$\overline{c}^i((x,y,w,\gamma),z) = \sum_{\ell\leq d} \frac{\partial}{\partial x_\ell}c^j(x,z)y^{\ell k};$$

- if $d+d^2<i\leq d+2d^2$ and i corresponds to the $(j,k)^{th}$ coordinate in $\mathbb{R}^d\otimes\mathbb{R}^d$:
  $$\overline{a}^i(x,y,w,\gamma) = \sum_{\ell\leq d} \frac{\partial}{\partial x_\ell}a^j(x)w^{\ell k} + \sum_{n\leq m} b^{jn}(x)f^{nk}(x,y) - \sum_{\ell\leq\beta}\int_E \frac{\partial}{\partial z_\ell}c^j(x,z)g^{\ell k}(x,y,z)dz$$
  $$\overline{b}^{in}(x,y,w,\gamma) = \sum_{\ell\leq d} \frac{\partial}{\partial x_\ell}b^{jn}(x)w^{\ell k}$$
  $$\overline{c}^i((x,y,w,\gamma),z) = \sum_{\ell\leq d} \frac{\partial}{\partial x_\ell}c^j(x,z)w^{\ell k} + \sum_{\ell\leq\beta} \frac{\partial}{\partial z_\ell}c^i(x,z)g^{\ell k}(x,y,z);$$

- if $d+2d^2<i\leq\overline{d}$ and $j=i-d-2d^2$:
  $$\overline{a}^i(x,y,w,\gamma) = 0, \quad \overline{b}^{in}(x,y,w,\gamma) = -f^{nj}(x,y),$$
  $$\overline{c}^i((x,y,w,\gamma),z) = \sum_{\ell\leq\beta} \frac{\partial}{\partial z_\ell}g^{\ell j}(x,y,z).$$

We wish to prove that $(\overline{a},\overline{b},\overline{c})$ satisfy $(\underline{A}'-(r-1))$. Our first task is to find a grading for $F=\mathbb{R}^{\overline{d}}$. We do as follows:

[α] We first decompose the first factor $\mathbb{R}^d$ according to $\mathbb{R}^{d_1}\times\ldots\times\mathbb{R}^{d_q}$.

[β] Next, we decompose the second factor $\mathbb{R}^d\otimes\mathbb{R}^d$ by arranging the coordinates $(i,j)_{i,j\leq d}$ into $q(q+1)/2$ blocks as follows: first $\{(i,j): i,j\leq M_1\}$; then $\{(i,j): i\leq M_1<j\leq M_2$ or $j\leq M_1<i\leq M_2\}$; then $\{(i,j): M_1<i,j\leq M_2\}$; then $\{(i,j): i\leq M_1$ and $M_2<j\leq M_3$, or $j\leq M_1$ and $M_2<i\leq M_3\}$; then $\{(i,j): M_1<i\leq M_2<j\leq M_3$ or $M_1<j\leq M_2<i\leq M_3\}$; then $\{(i,j): M_2<i,j\leq M_3\}$, and so on...

[γ] Next, we decompose the third factor $\mathbb{R}^d\otimes\mathbb{R}^d$ as above in [β].

[δ] Finally we decompose the fourth factor $\mathbb{R}^d$ as in [α].

So we have $\mathbb{R}^{\overline{d}} = \mathbb{R}^{d_1}\times\ldots\times\mathbb{R}^{d_{\overline{q}}}$ with $\overline{q}=2q+q(q+1)$. It is then tedious, but pretty obvious, to check that $(\overline{a},\overline{b},\overline{c})$ satisfies (A'-(r-1)) (use 6-38 and 6-39 and (A'-r) for (a,b,c)), with $\theta+1$ instead of θ.

Then $D\overline{X}^x$ exists by 6-24, and $\nabla\overline{X}^x$ exists by 6-29. Moreover, writing 6-25 for the relevant components of $D\overline{X}^x$ to get $D\nabla X^x$ or 6-30 for the relevant components of $\nabla\overline{X}^x$ to get ∇DX^x, give the same equation: so $D\nabla X^x=\nabla DX^x$.

Now we can iterate the procedure: $Z^{2,x}=(\overline{X}^x,\nabla\overline{X}^x,D\overline{X}^x)$ satisfies an equation 6-2 with (A'-(r-2)), then $Z^{3,x}=(Z^{2,x},\nabla Z^{2,x},DZ^{2,x})$ satisfies 6-2 with (A'-(r-3)),..., and we stop at $Z^{r-2,x}$, thus getting the result.

§6-f. INTEGRATION-BY-PARTS SETTING FOR X_t^x

As already pointed out in §6-a, Bismut's approach provides a natural integration-by-parts setting for every r.v. Φ that is F-(1)-differentiable (see 6-43). We restrict our attention to $\Phi=X_t^x$.

6-45 PROPOSITION: *Assume (A'-2). For all* $x\in\mathbb{R}^d$, $t\leq T$, *the following term* $(\sigma_t^x,\gamma_t^x,H_t^x,\delta_t^x)$ *is an integration-by-parts setting* (see 4-4) *for* X_t^x:

$$\left.\begin{aligned} \sigma_t^x &= DX_t^x, \quad \gamma_t^x = -\partial G_t^x, \\ H_t^x &= \text{the set of all } \underline{\underline{F}}_t\text{-measurable r.v. } \Psi \\ &\quad \text{such that } \{\Psi\circ\theta^{x,\lambda}\}_\lambda \text{ is F-differen-} \\ &\quad \text{tiable at } 0, \\ \delta_t^x(\Psi) &= -\partial\Psi^x, \text{ derivative at 0 of } \{\Psi\circ\theta^{x,\lambda}\}_{\lambda\in\Lambda}. \end{aligned}\right\} \quad (6\text{-}46)$$

<u>Proof</u>. A standard argument shows that if $\Psi\in(H_t^x)^n$ and $f\in C_p^2(R^n)$, then $f(\Psi)\in H_t^x$, and the derivative $\partial f(\Psi)^x$ of $\{f(\Psi)\circ\theta^{x,\lambda}\}$ at 0 is $\partial f(\Psi)^x=\sum_{i\leq n}(\partial f/\partial x_i)(\Psi)(\partial\Psi^i)^x$, so 4-5 holds. If $\Psi\in H_t^x$, 6-16 and the fact that $G^{x,\lambda}$ is a P-martingale show that

$$\begin{aligned} &E[\,f(X_t^x\circ\theta^{x,\lambda})\;\Psi\circ\theta^{x,\lambda}\;G_t^{x,\lambda}] \\ &\quad = E^{x,\lambda}[\,f(X_t^x\circ\theta^{x,\lambda})\;\Psi\circ\theta^{x,\lambda}] = E[\,f(X_t^x)\Psi]\,, \end{aligned}$$

for $f\in C_p^2(R^d)$. Differentiating this in λ at $\lambda=0$ yields 4-6 (recall that $G_t^{x,0}=1$).

In order to apply the results of Section 4, we still have to describe the sets $C_{t,0}^x(q)$ introduced in it. To do this, for each j such that $(A'-(j+1))$ holds we set:

$$Z^{j,x} = \begin{cases} (X^x, DX^x, \partial G^x) & \text{if } j=0 \\ (\{D^s\nabla^k X^x\}_{0\leq s+k\leq j}, \{D^s\nabla^k\partial G^x\}_{0\leq s+k\leq j-1}) & \text{if } j>1 \end{cases}$$

and for $C_{t,0}^x(q)$ we take the family of all components of $Z_t^{q,x}$: it contains in particular the components of σ_t^x, γ_t^x, $\nabla^i X_t^x$ for $0\leq i\leq q$, as was required. Then, if $Y_{t,\ell}^x(q)$ is the r.v. whose components constitute the set $C_{t,\ell}^x(q)$ (see §4-b), we get:

6-47 <u>LEMMA</u>: *Assume (A'-r) for some* $r\geq 3$ *and 6-38. Then a)* $\{X_t^x\}$ *is* (r-1) *times F-differentiable.*

b) $C^x_{t,r-2-q}(q) \subset H^x_t$ *for* $1 \leq q \leq r-2$, $C^x_{t,r-3}(0) \subset H^x_t$.

c) $x \rightsquigarrow \sup_{t \leq T} E(|Y^x_{n,t}(q)|^p)$ *is locally bounded for all* $p < \infty$, *provided* $n+q \leq r-1$ *if* $q \geq 1$, *and* $n \leq r-2$ *if* $q=0$.

Proof. (a) comes from 6-44. Note that $Z^{q,x}$ is the same as in the proof of 6-44 for $q \geq 2$, while $Z^{1,x} = \bar{X}^x$ and $Z^{0,x}$ is just a part of $Z^{1,x}$. So $Y^x_{t,\ell}(q)$ is a part of the components of $Z_t^{q+\ell,x}$ if $q \geq 1$, of $Z_t^{\ell+1,x}$ if $q=0$.

Now, in the proof of 6-44 we saw that $Z^{q,x}$ satisfies Equation 6-2 with (A'-(r-q)). So by 6-16, $Z_t^{q,x} \in H_t^x$ whenever $r-q \geq 2$. Moreover if $r-q \geq 1$, then $Z^{q,x}$ satisfies 6-2 as well with, at least, (A'-1). Then the estimate 5-11 gives

$$\|(Z^{q,x})^*_T\|_{L^p} \leq c_p(1+|x|),$$

since the initial condition $Z_0^{q,x}$ is $(x,I,0,0)$. Hence we have (b) and (c).

Then the results of Section 4 immediately yield:

6-48 THEOREM: *Assume (A'-j) and 6-38, and set* $Q_t^x = \det(DX_t^x)$ *and*

$$q_t^x(i) = E(|Q_t^x|^{-i}) \qquad (=\infty \quad \text{if } P(Q_t^x=0)>0). \tag{6-49}$$

a) If $j \geq 3$ *and* $Q_t^x \neq 0$ *a.s.,* X_t^x *admits a density* $y \rightsquigarrow p_t(x,y)$.

b) Moreover, $p_t(x,.)$ *is of class* C^r, *provided:*
- *either* $j \geq r+d+3$ *and* $q_t^x(2r+2d+2+\varepsilon) < \infty$ *for some* $\varepsilon > 0$,
- *or* $j \geq r+3$ *and* $q_t^x(2d(r+1)+\varepsilon) < \infty$ *for some* $\varepsilon > 0$.

c) Moreover, $(x,y) \rightsquigarrow p_t(x,y)$ *is of class* C^r, *provided:*

- *either* $j \geq r+2d+3$ *and* $x \rightsquigarrow q_t^x(2r+4d+2+\varepsilon)$ *is locally bounded for some* $\varepsilon>0$.
- *or* $j \geq r+3$ *and* $x \rightsquigarrow q_t^x(4d(r+1)+\varepsilon)$ *is locally bounded for some* $\varepsilon>0$.

d) Moreover:

(i) If $j \geq 2r+4d+6$, *if for every bounded subset* $A \subset \mathbb{R}^d$, $\sup_{t>t_0, x\in A} q_t^x(4r+8d+8+\varepsilon)<\infty$ *for some* $\varepsilon>0$, *and if*

$$|\det[I + v D_x c(x,z)]| \geq \zeta \quad \forall v \in [0,1] \tag{6-50}$$

for some constant $\zeta>0$, *then* $(t,x,y) \rightsquigarrow p_t(x,y)$ *is of class* C^r *on* $(t_0,T] \times \mathbb{R}^d \times \mathbb{R}^d$.

(ii) If $j \geq 2r+4$, *if for every bounded subset* $A \subset \mathbb{R}^d$, $\sup_{t>t_0, x\in A} q_t^x(4(r+1)(2d+1)+\varepsilon)<\infty$ *for some* $\varepsilon>0$, *and if* $c \equiv 0$ (i.e., the Poisson measure in 6-2 does not intervene), *then* $(t,x,y) \rightsquigarrow p_t(x,y)$ *is of class* C^r *on* $(t_0,T] \times \mathbb{R}^d \times \mathbb{R}^d$.

Proof. (a), (b), (c) immediately follow from Theorems 4-7, 4-19, 4-21, and Lemma 6-47. (d) also follows from this lemma and from Theorem 4-31, provided we prove the next result.

6-51 LEMMA: *Properties 6-50 and (A'-j) for some* $j \geq 2$ *imply that the family* $\{\Phi_t^x = X_t^x\}$ *meets 4-26 for* j, *and if* $c \equiv 0$ *we have* $\phi_u(x)=x$ *in 4-26.*

Proof. Recall first that the extended generator of the Markov process X^x (see [13]) operates on $C_p^2(R^d)$ as

$$\begin{aligned} \mathcal{L}f(x) &= a(x)D_x f(x) + \tfrac{1}{2}bb^T(x)\, D_{x^2}^2 f(x) \\ &\quad + \int_E [f(x+c(x,z))-f(x)-D_x f(x)c(x,z)]\,dz \end{aligned} \tag{6-52}$$

and $f(X_t^x)-f(x)-\int_0^t Lf(X_s^x)ds$ is a local martingale. Now, under (A'-1), $|X^x|_T^* \in \cap_{p<\infty} L^p$, while $|Lf(x)| \leq \zeta(1+|x|^\theta)$ for suitable ζ, θ. Hence $|Lf(X^x)|_T^* \in \cap_{p<\infty} L^p$, and the above local martingale is a martingale, and

$$E(f(X_t^x)) = f(x) + \int_0^t E(Lf(X_s^x))ds.$$

Since f is continuous, $s \rightsquigarrow E(Lf(X_s^x))$ is continuous, and the above yields that $t \rightsquigarrow E(f(X_t^x))$ is differentiable, with derivative $E(Lf(X_t^x))$.

So it remains to prove that f is of the form 4-28, with m, A^i, ϕ_u satisfying (i)-(iii) of 4-26. Second order Taylor's formula with integral rest yields

$$\begin{aligned} & f(x+c(x,z))-f(x)-D_x f(x)c(x,z) \\ & = \int_0^1 (1-v) \langle D^2_{x^2} f(x+vc(x,z)), c(x,z)\otimes c(x,z)\rangle \, dv, \end{aligned}$$

so that 6-52 is exactly 4-28, provided we put (η being the auxiliary function in (A'-j)):

$$U = \{0\}\cup(E\times[0,1]),$$

$$m = \varepsilon_0 + \eta^2(z)G(dz)\times(1-v)1_{[0,1]}(v)dv,$$

$$A_u^1 = a, \quad A_u^2 = \tfrac{1}{2}bb^T, \quad \phi_u(x) = x \qquad \text{if } u=0,$$

$$A_u^1 = 0, \quad A_u^2(x) = c(x,z)\otimes c(x,z)/\eta^2(z),$$

$$\text{and} \quad \phi_u(x) = x+vc(x,z) \text{ if } u=(z,v)\ E\times[0,1].$$

The regularity assumptions in 4-26 are then clearly met by these terms (recall that the function η is in $L^2(E,G)$).

Finally, the last claim is evident, with the above definition of ϕ_u.

Section 7: PROOF OF THE MAIN THEOREMS VIA BISMUT'S APPROACH

§7-a. INTRODUCTORY REMARKS

Our aim now is to deduce Theorems 2-14, 2-27, 2-28, 2-29 from Theorem 6-48. However, the settings in Section 2 and in Section 6 are different, and we first have to conciliate them both. This is presently achieved, through the following remarks.

1) The driving terms are an m-dimensional Brownian motion W and A+1 Poisson measures $\mu_1,\ldots,\mu_A$ and μ, as in 2-1. It is no restriction for the results of Section 2 to assume that the filtered space supporting W,μ_α,μ is the *canonical space*, as defined in §6-a, but with A+1 Poisson random measures instead of one.

2) As mentionned in Remark 2-16, it is possible to weaken Assumptions (A-r) into (A'-r), for the coefficients of Equation 2-2. However, the space E on which the last Poisson measure μ sits is an "abstract space", and there is no differentiability in z for the coefficient c. Hence, one should state Assumption (A'-r) as follows:

7-1 ASSUMPTION (A'-r): a,b, *and each* c_α, *meet 6-3 (the latter with some* $\eta_\alpha\in\cap_{2\leq p<\infty}L^p(E_\alpha,\underline{\underline{E}}_\alpha)$). *The coefficient* c *meets (i) and (iii) of 6-3 with some* $\eta\in\cap_{2\leq p<\infty}L^p(E,G)$.

Then obviously (A-r) ⇒ (A'-r).

3) All results of Section 5 are of course valid with $A+1$ Poisson measures instead of one.

4) Now we describe the changes to be undergone in Section 6. We will perturb *each measure* μ_α, *but not* μ.

So for each $\alpha=1,\dots,A$ we start with a function ρ_α: $E_\alpha \to [0,\infty)$ meeting 6-9, and with a function v_α meeting 6-10 relatively to ρ_α. Then Equation 6-15 reads

$$G^\lambda = 1 - [G^\lambda_-(u.\lambda)]*W + \sum_\alpha [G^\lambda_-(Y^\lambda_\alpha - 1)]*\tilde{\mu}_\alpha .$$

The reader may also think that one "perturbs" μ as well, but with the function $v\equiv 0$! so all of Section 6 goes through without modification, except that every integral with respect to μ (resp. $\tilde{\mu}$) should be replaced by a sum of integrals with respect to μ_α and μ (resp. $\tilde{\mu}_\alpha$ and $\tilde{\mu}$), with the convention $v\equiv 0$.

In particular, ∇X^x is still given by 6-32, with 6-31 replaced by

$$\begin{aligned} K^x &= D_x a(X^x_-)*t + D_x b(X^x_-)*W \\ &\quad + \sum_\alpha D_x c_\alpha(X^x_-)*\tilde{\mu}_\alpha + D_x c(X^x_-)*\tilde{\mu}. \end{aligned} \tag{7-2}$$

6-34 also holds, with (instead of 6-41):

$$H^x = b(X^x_-)f(X^x_-,\nabla X^x_-)*t + \sum_\alpha D_z c_\alpha(X^x_-)g_\alpha(X^x_-,\nabla X^x_-)*\mu_\alpha \tag{7-3}$$

with g_α satisfying 6-38-(ii) relatively to ρ_α. The fundamental formulae 6-35, 6-36, 6-37 remain unchanged, as well of course as Theorems 6-44 and 6-48.

§7-b. EXISTENCE OF THE DENSITY

We begin with auxiliary results.

7-4 LEMMA: *Under (A'-1), the following* (where B, C_α are defined in 2-10)

$$R^x = B(X^x_-)*t + \sum_\alpha C_\alpha(X^x_-)*\mu_\alpha \tag{7-5}$$

defines a process with values in the set of all symmetric nonnegative $d\times d$ *matrices, and this process is nondecreasing for the strong order in this set.*

Proof. Since B and C_α are nonnegative symmetric matrices, and $\rho_\alpha \geq 0$, it is clearly enough to prove that R^x_t is finite-valued for all $t \leq T$.

In view of (A'-1), $B(X^x_-)$ is locally bounded and so $B(X^x_-)*t$ is finite-valued. In view of 2-10 and of (A'-1) again,

$$|C_\alpha(x,z)| \leq \zeta'(1+|x|^{\theta'})(1+\eta'_\alpha(z)) \\ g_\alpha(x,z)^{-2}\, 1_{\{g_\alpha(x,z)\neq 0\}} \tag{7-6}$$

for some constants ζ', θ' and $\eta'_\alpha \in \cap_{2\leq p<\infty} L^p(E_\alpha, G_\alpha)$, and $g_\alpha = |\det(I+D_x c_\alpha)|$. Since $\rho_\alpha \in L^1(E_\alpha, G_\alpha)$, we have a.s.

$$[(1+|X^x_-|^{\theta'})(1+\eta'_\alpha)\rho_\alpha * \mu_\alpha]_T < \infty.$$

Hence, due to 7-6, it remains to show that for almost all ω,

$$g_\alpha(X^x_{t-}(\omega),z) \geq \gamma(\omega), \mu_\alpha(\omega;dt,dz)\text{-a.s.} \\ \text{on the set } \{(t,z): g_\alpha(X^x_{t-}(\omega),z)\neq 0\}, \tag{7-7}$$

where $\gamma(\omega)>0$. Let $\tilde{g}_t(\omega)=\det(I+\Delta K^x_t(\omega))$. In view of 7-2, outside a P-null set, 7-7 is equivalent to saying that

$$\text{for all } t\leq T, \text{ either } |\tilde{g}_t(\omega)|\geq\gamma(\omega), \text{ or } \tilde{g}_t(\omega)=0. \tag{7-8}$$

Now, $\nabla X^x(n) = \nabla X^x(n)_-(I+\Delta K^x(n))$ by 6-35. Hence for

$T_n^x < t < T_{n+1}^x$, $\det(\nabla X^x(n)_t) = \det(\nabla X^x(n)_{t-})\tilde{g}_t(\omega)$. In virtue of 6-36, $|\det(\nabla X^x(n))|$ is bounded, and bounded away from 0 (with bounds depending on ω) on the interval $[0, T_{n+1}^x)$. Hence there exists $\gamma_n(\omega) > 0$ such that

$$|\tilde{g}_t(\omega)| \geq \gamma_n(\omega) \qquad \text{for} \quad T_n^x(\omega) < t < T_{n+1}^x(\omega).$$

Moreover $\tilde{g}_t = 0$ if $t = T_n^x$. Thus 7-8 is met with $\gamma(\omega) = \inf_{n \leq N(\omega)} \gamma_n(\omega)$, where $N(\omega)$ is such that $T_{N(\omega)}^x(\omega) \leq T < T_{N(\omega)+1}^x(\omega)$; since $N(\omega) < \infty$, we have $\gamma(\omega) > 0$ and the proof is finished.

7-9 LEMMA: *Assume (A'-1) and (B)* (see 2-13), *and that* $\rho_\alpha > 0$ *on* E_α *for all* $\alpha = 1, \ldots, A$. *Then* P-*a.s. the matrices* $R_t^x - R_s^x$ *are invertible for all* $0 \leq s < t \leq T$.

Proof. For simplicity, we omit the superscript "x". Let $s < T$ be fixed. Let Y be a unit random vector in $\mathbb{R}^d$, which is $\underline{\underline{F}}_s$-measurable. With Γ_α as in (B), we set $\Gamma'_\alpha = \{(x,z) \in \Gamma_\alpha : G_\alpha(\Gamma_{\alpha,x}) = +\infty\}$, which is Borel, and

$$V^n = \{(\omega,t) : t > s, Y^T(\omega) B(X_{t-}(\omega)) Y(\omega) \geq \tfrac{1}{n}\},$$

$$V_\alpha^n = \{(\omega,t,z) : t > s, (X_{t-}(\omega), z) \in \Gamma'_\alpha, \\ Y^T(\omega) C_\alpha(X_{t-}(\omega), z) Y(\omega) \geq \tfrac{1}{n}\},$$

$$V = \lim_n \uparrow V^n, \qquad V_\alpha = \lim_n \uparrow V_\alpha^n,$$

$$Z^n = 1_{V^n} * t + \sum_\alpha \rho_\alpha 1_{V_\alpha^n} * \mu_\alpha,$$

$$Z = 1_V * t + \sum_\alpha \rho_\alpha 1_{V_\alpha} * \mu_\alpha.$$

As $\rho_\alpha \in L^1(E_\alpha, G_\alpha)$, Z^n and Z are finite-valued increasing processes, and Z_t^n increases to Z_t as $n \uparrow \infty$. Moreover 7-5 yields for $t > s$:

$$Y^T R_t Y - Y^T R_s Y \geq (Y^T B(X_-) Y 1_{V^n} * t)_t + \sum_\alpha (Y^T C_\alpha(X_-) Y \rho_\alpha 1_{V_\alpha^n} * \mu_\alpha)_t \geq \frac{1}{n} Z_t^n. \tag{7-10}$$

Let $\theta>0$ and $U^\theta = e^{-\theta Z}$. Then Ito's formula for processes with finite variation yields

$$\begin{aligned} U^\theta &= 1 - \theta U_-^\theta 1_V * t - \sum_\alpha \theta \rho_\alpha U_-^\theta 1_{V_\alpha} * \mu_\alpha \\ &\quad + \sum_\alpha U_-^\theta (e^{-\theta\rho_\alpha} - 1 + \theta\rho_\alpha) 1_{V_\alpha} * \mu_\alpha \\ &= 1 - \theta U_-^\theta 1_V * t - \sum_\alpha U_-^\theta (1 - e^{-\theta\rho_\alpha}) 1_{V_\alpha} * \mu_\alpha. \end{aligned} \tag{7-11}$$

Now $0 \leq U^\theta \leq 1$ and $0 \leq 1 - e^{-\theta\rho_\alpha} \leq \theta\rho_\alpha$, which is G_α-integrable. So when taking expectationsin 7-11 we may replace μ_α by ν_α (noting that V_α is predictable). We obtain for $t>s$:

$$\begin{aligned} E(U_t^\theta) &= 1 - \int_s^t E\{\theta U_r^\theta 1_V(r) \\ &\quad + \sum_\alpha U_r^\theta \int_{E_\alpha} (1 - e^{-\theta\rho_\alpha(z)}) 1_{V_\alpha}(r,z) dz\} dr \\ &\leq 1 - \int_s^t E\{\theta 1_V(r) 1_{\{Z_r=0\}} \\ &\quad + \sum_\alpha 1_{\{Z_r=0\}} \int_{E_\alpha} (1 - e^{-\theta\rho_\alpha(z)}) 1_{V_\alpha}(r,z) dz\} dr. \end{aligned}$$

Each individual term above is increasing in θ, so we may pass to the limit as $\theta \uparrow \infty$. With the convention $0 \times \infty = 0$, we get (because $\rho_\alpha > 0$):

$$\begin{aligned} P(Z_t=0) \leq 1 - \int_s^t E\{+\infty\ 1_V(r) 1_{\{Z_r=0\}} \\ + \sum_\alpha 1_{\{Z_r=0\}} \int_{E_\alpha} 1_{V_\alpha}(r,z) dz\} dr. \end{aligned} \tag{7-12}$$

In the following, we use the notation W_x^α, N_x, N_{xz}^α of 2-11 and 2-13. We have $V = \{(\omega,t): t>s, Y(\omega) \notin N_{X_{t-}(\omega)}\}$ (recall that $V = \bigcup_n V^n$), so if

$V'_\alpha := \{(\omega,t): t>s, Y(\omega) \notin W^\alpha_{X_{t-}(\omega)}\}$, Assumption (B) yields

$$V \cup (\bigcup_\alpha V'_\alpha) = \Omega \times (s,T]. \qquad (7\text{-}13)$$

Since $V_\alpha = \bigcup_n V^n_\alpha$, we also have

$$V_\alpha = \{(\omega,t,z): t>s, z \in \Gamma'_{\alpha,X_{t-}(\omega)}, Y(\omega) \notin N^\alpha_{X_{t-}(\omega),z}\}.$$

Therefore if $t>s$ and $Y(\omega) \notin W^\alpha_{X_{t-}(\omega)}$ we have $\Gamma'_{\alpha,X_{t-}(\alpha)} =$ $\{z:(\omega,t,z) \in V_\alpha\}$, and thus (recalling 2-13):

$$\int_{E_\alpha} 1_{V_\alpha}(\omega,t,z)dz \geq \infty \times 1_{V'_\alpha}(\omega,t)$$

for all ω,t such that $Y(\omega) \notin W^\alpha_{X_{t-}(\omega)}$, $t>s$. Hence 7-13 yields for $r>s$:

$$\begin{aligned}\infty 1_V(\omega,r) &+ \sum_\alpha \int_{E_\alpha} 1_{V_\alpha}(\omega,s,z)dz \\ &\geq \infty 1_V(\omega,r) + \infty \times (\sum_\alpha 1_{V'_\alpha}(\omega,r)) = \infty\end{aligned}$$

(note that we have not used the (usually wrong) fact that $V'_\alpha \in \underline{\underline{F}} \times \underline{\underline{B}}([0,T])$). Substituting in 7-12, we get for $t>s$:

$$P(Z_t=0) \leq 1 - \int_s^t E(\infty 1_{\{Z_r=0\}})dr.$$

This is only possible if $P(Z_r=0)=0$ for almost all r in (s,t). Since Z is increasing, we deduce that $Z>0$ a.s. on $(s,T]$.

Finally, consider 7-10 and recall that $Z^n \uparrow Z$. If $t>s$ we have $Z_t>0$ a.s., so $Z^n_t>0$ for n large enough, so $Y^T R_t Y > Y^T R_s Y$ a.s. Then, for almost all ω, $Y(\omega)$ does not belong to the kernel $L_t(\omega)$ of the linear map $R_t(\omega)-R_s(\omega)$. Because of 7-4, $L_t(\omega)$ increases as t decreases, and we set $L_{s+}(\omega) = \bigcup_{t>s} L_t(\omega)$. Then we have that $Y(\omega) \notin L_{s+}(\omega)$ for almost all ω. Since L_{s+} is $\underline{\underline{F}}_s$-measurable (in an obvious sense) and since the above property holds for each unit random vector Y that is $\underline{\underline{F}}_s$-measurable, we deduce that $L_{s+}(\omega)=\{0\}$ for almost all ω: this implies that R_t-R_s is a.s. invertible for all $t>s$.

Finally, let Ω_0 be the set where $R_t(\omega)-R_s(\omega)$ is invertible for all $s<t$, with $s,t\in\mathbb{Q}\cap[0,T]$. By what precedes, $P(\Omega_0)=1$. Let $\omega\in\Omega_0$ and $0\leq s<t\leq T$. There exist $r,r'\in\mathbb{Q}$ with $s\leq r<r'\leq t$ and $R_{r'}(\omega)-R_r(\omega)$ is invertible. Because R is increasing for the strong order (see 7-4), $R_t(\omega)-R_s(\omega)$ is a-fortiori invertible, and the claim is proved.

For the next result, we recall that K^x is defined in 7-2, and that $T_1^x=\inf(t:\det(I+\Delta K_t^x)=0)$ (see 6-35).

7-14 <u>LEMMA</u>: *Assume (A'-3) and (B). Then if* $t\in(0,T]$, *the law of* X_t^x *under the restriction of the probability to the set* $\{t<T_1^x\}$ *admits a density* :in other words, if A is a Borel subset of $\mathbb{R}^d$ with Lebesgue measure 0,

$$P(X_t^x\in A, t<T_1^x) = 0 \tag{7-15}$$

<u>Proof</u>. We will apply the following extension of Theorem 6-48-a: in view of 6-45 and 6-47 and of the first part of Theorem 4-7, and since (A'-3) holds, the claim will be proved if we can choose u and v_α so that

a) 6-38 holds, and

b) the matrix DX_t^x is a.s. invertible on the set $\{t<T_1^x\}$.

For proving (b) we will use Lemma 7-9, which requires $\rho_\alpha>0$ on E_α. Since E_α is a countable union of β_α-dimensional rectangles, it is easy to find a function ρ_α which meets both 6-9 and $\rho_\alpha>0$ on E_α.

The choice of u and v_α amounts to choosing functions f and g_α which meet 6-38, and it intervenes in the expression 7-3 for the process H^x. Now, 6-37 yields

$$DX_t^x = \nabla X_t^x \int_0^t (\nabla X_{s-}^x)^{-1}(I+\Delta K_s^x)^{-1} dH_s^x \quad \text{if } t<T_1^x,$$

while ∇X_t^x is invertible for $t<T_1^x$. So (b) amounts to

$$\int_0^t (\nabla X_{s-}^x)^{-1}(I+\Delta K_s^x)^{-1} dH_s^x \text{ is a.s. invertible on } \{t<T_1^x\}. \tag{7-16}$$

Suppose for a moment that for f and g_α we could choose the following functions $\bar{f}$ and $\bar{g}_\alpha$ (here, $x\in\mathbb{R}^d$ and $y\in\mathbb{R}^d\otimes\mathbb{R}^d$ and $z\in E_\alpha$):

$$\left.\begin{array}{l} \bar{f}(x,y) = b(x)^T y^{-1,T} \quad (=0 \text{ if } y \text{ is not invertible}) \\ \bar{g}_\alpha(x,y,z) = D_z c_\alpha(x,z)^T (I+D_x c_\alpha(x,z))^{-1,T} y^{-1,T} \\ \quad (=0 \text{ if } y \text{ or } I+D_x c_\alpha \text{ are not invertible}). \end{array}\right\} \tag{7-17}$$

Then the process in 7-16 becomes, in view of 7-5:

$$\int_0^t (\nabla X_{s-}^x)^{-1}\, dR_s^x\, (\nabla X_{s-}^x)^{-1,T} \quad \text{for } t<T_1^x,$$

which in virtue of Lemma 7-9 is clearly invertible a.s.

Now, the functions 7-17 do not meet 6-38. To overcome this little difficulty, we proceed as follows. We define the following functions:

$$\left.\begin{array}{l} h(y)=(1+|y|^2)^{-1}: \mathbb{R}^d\otimes\mathbb{R}^m \to (0,1] \\ h'_\alpha(y)=(1+|y|^2)^{-1}: \mathbb{R}^d\otimes\mathbb{R}^{\beta_\alpha} \to (0,1] \\ h'': \mathbb{R}^d\otimes\mathbb{R}^d \to [0,1] \text{ is such that } h''(y)=0 \text{ if} \\ \quad \text{and only if } y \text{ is not invertible, and} \\ \quad y \rightsquigarrow y^{-1,T}h''(y) \text{ is of class } C_b^\infty \text{ on } \mathbb{R}^d\otimes\mathbb{R}^d \end{array}\right\} \tag{7-18}$$

(there are plenty of such choices for h"). Then $\bar{f}$ and

$\bar{g}_\alpha$ being given again by 7-17, we set:

$$\left.\begin{aligned} f(x,y) &= \bar{f}(x,y)h(b(x))h''(y) \\ g_\alpha(x,y,z) &= \bar{g}_\alpha(x,y,z)h'_\alpha(D_z c_\alpha(x,z)) \\ &\qquad h''(I+D_x c_\alpha(x,z))h''(y). \end{aligned}\right\} \quad (7\text{-}19)$$

It is then easy to check that f and g_α meet 6-38. Moreover, the process in 7-16 reads as

$$\int_0^t (\nabla X^x_{s-})^{-1}\, d\hat{R}^x_s\, (\nabla X^x_{s-})^{-1,T} \quad \text{for } t<T^x_1 \qquad (7\text{-}20)$$

where

$$\begin{aligned} \hat{R}^x &= B(X^x_-)h(b(X^x_-))h''(\nabla X^x_-)*t \\ &+ \sum_\alpha C_\alpha(X^x_-)\rho_\alpha h'_\alpha(D_z c_\alpha(X^x_-))h''(I+D_x c_\alpha(X^x_-))h''(\nabla X^x_-)*\mu_\alpha. \end{aligned} \qquad (7\text{-}21)$$

Since $h>0$, $h'_\alpha>0$, and $h''(y)>0$ for y invertible, comparing 7-21 and 7-5 yields that $\hat{R}^x_t-\hat{R}^x_s$ is invertible whenever $R^x_t-R^x_s$ is so, and the latter is true a.s. when $t>s$ by Lemma 7-9. It is then obvious to deduce (using 7-20) that 7-16 holds, and we are finished.

7-22 Proof of Theorem 2-14. We will deduce the existence of a density for X^x_t (where $t\in(0,T]$) from Lemma 7-14, via a Markov process argument which has already been used in this context by Léandre [18]. It is well known that the solution to 2-2 is strong Markov; however, in order to properly apply the strong Markov property at time T^x_n we need to be somewhat careful.

Firstly, we can indeed assume that the time interval is $\mathbb{R}_+$ instead of $[0,T]$; this slightly simplifies the description of the shift semi-group, which goes as follows: θ_t is the unique map from the canonical space Ω into itself, such that $W_{t+s}-W_s=W_s\circ\theta_t$ and

$\mu(\omega;(t+ds),dz)=\mu(\theta_t\omega,ds,dz)$, and similarly for all μ_α. The independence and stationarity of the increments of W,μ_α,μ then yield that $P\circ\theta_S^{-1}=P$ for every finite stopping time S.

Secondly, let S be a finite stopping time; then $W\circ\theta_S$ is again a Brownian motion, and we have

$$\int_S^{S+t} b(X_s^x)\,dW_s = \int_0^t b(X_{S+u}^x)\,dW_u\circ\theta_S$$

up to a null set. Similar equalities hold for the other terms appearing in Equation 2-2, and thus

$$X_{S+t}^x = X_S^x + a(X_{(S+.)-}^x)*t + b(X_{(S+.)-}^x)*(W\circ\theta_S) + \sum_\alpha c_\alpha(X_{(S+.)-}^x)*(\tilde{\mu}_\alpha\circ\theta_S) + c(X_{(S+.)-}^x)*(\tilde{\mu}\circ\theta_S)$$

up to a null set. So the uniqueness of the solution to 2-2 yields

$$X_{S+t}^x = X_t^Y\circ\theta_S \quad \text{a.s. for all } t, \text{ where } Y=X_S^x. \tag{7-23}$$

The same argument applied to 7-2 also gives

$$K_{S+t}^x = K_S^x + K_t^Y\circ\theta_S \quad \text{a.s. for all } t, \text{ with } Y=X_S^x. \tag{7-24}$$

Next, we will apply this to the stopping time $S=T_n^x$ defined in 6-35, and we put again $Y=X_S^x$. Then 6-35 and 7-24 yield

$$T_{n+1}^x = T_n^x + T_1^Y\circ\theta_{T_n^x} \quad \text{a.s. on } \{T_n^x<\infty\}. \tag{7-25}$$

We have seen that $P\circ\theta_S^{-1}=P$, and $\theta_S^{-1}(\underline{\underline{F}})$ is independent from $\underline{\underline{F}}_S$ on $\{S<\infty\}$. Then, if A is a Borel subset of $\mathbb{R}^d$, 7-23 and 7-25 imply

$$P(X_t^x\in A, T_n^x<t<T_{n+1}^x) = P(T_n^x<t, X_{t-T_n^x}^Y\circ\theta_{T_n^x}\in A, T_1^Y\circ\theta_{T_n^x}>t-T_n^x)$$

$$= \int P(d\omega) 1_{\{T_n^x(\omega)<t\}} P(X^{Y(\omega)}_{t-T_n^x(\omega)} \in A, T_1^{Y(\omega)} > t - T_n^x(\omega)).$$

If moreover A has Lebesgue measure 0, we deduce from 7-15 that

$$P(X_t^x \in A, T_n^x < t < T_{n+1}^x) = 0. \tag{7-26}$$

Finally, $P(T_n^x = t) = 0$ for all n, because T_n^x is one of the jump times of one of the Poisson measures μ_α or μ. So summing up 7-26 on n gives $P(X_t^x \in A) = 0$, which is exactly saying that X_t^x has a density.

§7-c. SMOOTHNESS OF THE DENSITY

Again, we begin with auxiliary results.

7-27 LEMMA: *Under (A'-2) and (SC) (see 2-26), for all* $p<\infty$ *the function* $x \rightsquigarrow E(|(\nabla X^x)^{-1}|_T^{*p})$ *is locally bounded.*

Proof. Recalling that

$$\nabla X^x = I + D_x a(X_-^x)\nabla X_-^x * t + D_x b(X_-^x)\nabla X_-^x * W$$
$$+ \sum_\alpha D_x c_\alpha(X_-^x)\nabla X_-^x * \tilde{\mu}_\alpha + D_x c(X_-^x)\nabla X_-^x * \tilde{\mu},$$

it is well known that $Y^x = (\nabla X^x)^{-1}$ is the solution to

$$Y^x = I + Y_-^x A^x * t + Y_-^x B^x * W + \sum_\alpha Y_-^x C_\alpha^x * \tilde{\mu}_\alpha + Y_-^x C^x * \tilde{\mu} \tag{7-28}$$

where, due to (SC), we can define

$$A_t^x = -D_x a(X_{t-}^x) + \sum_{i=1}^m D_x b^{\cdot i}(D_x b^{\cdot i})^T(X_{t-}^x)$$
$$+ \int_E [D_x c(I+D_x c)^{-1} D_x c](X_{t-}^x, z)G(dz)$$
$$+ \sum_\alpha \int_{E_\alpha} [D_x c_\alpha (I+D_x c_\alpha)^{-1} D_x c_\alpha](X_{t-}^x, z)dz$$

$$B_t^x = -D_x b(X_{t-}^x),$$
$$C_\alpha^x(t,z) = -D_x c_\alpha(X_{t-}^x,z)[I+D_x c_\alpha(X_{t-}^x,z)]^{-1},$$
$$C^x(t,z) = -D_x c(X_{t-}^x,z)[I+D_x c(X_{t-}^x,z)]^{-1}.$$

(to verify this claim, it suffices to define Y^x as the solution to 7-28, and to apply Ito's formula to the product $\tilde{Y}^x = \nabla X^x Y^x$: one obtains that $\tilde{Y}^x$ is the solution to a linear equation, to which the constant process I is also trivially a solution: hence $\tilde{Y}^x = I$).

Now, Equation 7-28 is of type 5-3. If $\mathbb{R}^d = \mathbb{R}^{d_1}\times\ldots\times\mathbb{R}^{d_q}$ is the grading in (A'-2), it is easy to recognize that the coefficients of 7-28 are graded, according to the grading of $\mathbb{R}^d\otimes\mathbb{R}^d$ described in [β] of the proof of 6-44, and due to (A'-2) and (SC) they satisfy 5-9, with Z_t^x and $\hat{Z}_t^x$ as follows:

$$Z_t^x = 0, \qquad \hat{Z}_t^x = \zeta(1+|X_{t-}^x|^\theta)$$

for some constants $\zeta,\theta\geq 0$. Since $\|(X^x)_T^*\|_{L^p}$ is locally bounded in x for every $p<\infty$, it is easily deduced from the estimate 5-11 that $x \rightsquigarrow \|(Y^x)_T^*\|_{L^p}$ is locally bounded as well.

7-29 <u>LEMMA</u>: *For every symmetric positive* $d\times d$ *matrix* A *and every* $p\in(0,\infty)$:

$$\frac{\Gamma(p)^d}{\det(A)^p} \leq \int_{\mathbb{R}^d} |x|^{d(2p-1)}\, e^{-x^T A x} dx \leq \frac{\Gamma(p)^d}{\det(A)^p} d^{d(2p-1)/2}$$

(where Γ is the usual gamma function).

<u>Proof</u>. There is an orthogonal transformation of $\mathbb{R}^d$ into itself that changes neither $\det(A)$, $|x|$, nor $x^T Ax$, but transforms A into a diagonal matrix. Hence, we may suppose that A is diagonal, with entries $\lambda_1>0,\ldots,$

$\lambda_d>0$. A simple substitution gives

$$\Gamma(p) = \lambda^p \int_{-\infty}^{+\infty} |z|^{2p-1} e^{-\lambda z^2} dz.$$

Hence

$$\frac{\Gamma(p)^d}{\det(A)^p} = \prod_{i=1}^d \frac{\Gamma(p)}{\lambda_i^p} = \prod_{i=1}^d \int_{-\infty}^{+\infty} |x_i|^{2p-1} e^{-\lambda_i (x_i)^2} dx_i$$

$$= \int_{\mathbb{R}^d} (\prod_{i=1}^d |x_i|)^{2p-1} e^{-x^T A x} dx$$

$$= \int_0^\infty r^{d-1} dr \int_{S_{d-1}} r^{d(2p-1)} (\prod_{i=1}^d \sigma_i)^{2p-1} e^{-r^2 \sigma^T A \sigma} d\sigma$$

$$\leq \int_0^\infty r^{2pd-1} dr \int_{S_{d-1}} e^{-r^2 \sigma^T A \sigma} d\sigma \qquad (7\text{-}30)$$

(since $|\sigma|=1$ if $\sigma \in S_{d-1}$, the d-dimensional unit sphere, so $|\sigma_i| \leq 1$). Finally, 7-30 equals

$$\int_{\mathbb{R}^d} |x|^{d(2p-1)} e^{-x^T A x} dx,$$

and the first inequality is proved. To prove the converse, we notice first that the function $f(\sigma) = \prod_{1 \leq i \leq d} |\sigma_i|$ takes its minimum on S_{d-1} in those points whose coordinates have moduli all equal to $d^{-1/2}$: thus $f(\sigma) \geq d^{-d/2}$ on S_{d-1}. Then the inequality 7-30 may be reversed, provided we multiply the right-hand side by $d^{-d(2p-1)/2}$.

7-31 <u>LEMMA</u>: *Assume (A'-3) and (SC) and (SB-(ζ,θ))* (see 2-24) *for some ζ,θ>0, and let ρ_α, ε, δ be like in (SB-(ζ,θ)); set*

$$S^x = k(X^x_-, \nabla X^x_-)(\nabla X^x_-)^{-1} B(X^x_-)(\nabla X^x_-)^{-1,T} * t$$
$$+ \sum_\alpha k'_\alpha(X^x_-, \nabla X^x_-, z)(\nabla X^x_-)^{-1} C_\alpha(X^x_-, z)(\nabla X^x_-)^{-1,T} \rho_\alpha(z) * \mu_\alpha \tag{7-32}$$

where $k: \mathbb{R}^d \times (\mathbb{R}^d \otimes \mathbb{R}^d) \to (0,1]$ *and* k'_α: $\mathbb{R}^d \times (\mathbb{R}^d \otimes \mathbb{R}^d) \times E_\alpha \to (0,1]$ *are Borel and satisfy for some constants* $\delta', \delta'' > 0$:

$$k(x,y),\ k'_\alpha(x,y,z) \geq \delta' \frac{|y|^{\delta''}}{(1+|x|^{\delta''})(1+|y|^{\delta''})}. \tag{7-33}$$

Then for all $t \in (0,T]$, $p \in (0,\infty)$, $n \in \mathbb{N}^*$ *such that* $pd < \zeta$ *and* $\frac{t}{n} \geq \theta$,

$$x \rightsquigarrow E(\det(S^x_t)^{-np}) \quad \text{is locally bounded.} \tag{7-34}$$

Proof. Since k, k'_α are bounded and $(\nabla X^x_-)^{-1}$ is locally bounded (see 7-27), if we compare 7-32 and 7-5 we observe that S^x is a process taking values in the set of $d \times d$ symmetric nonnegative matrices and is non-decreasing for the strong order in this set.

We now break the proof into several steps.

Step 1. We have (SB-(ζ,θ)) for some broad functions f_α. For $x, y \in \mathbb{R}^d$ set

$$\xi_\alpha(x,y) = \inf_{z: f_\alpha(z) > 0} \frac{\rho_\alpha(z)}{f_\alpha(z)}\, y^T C_\alpha(x,z) y.$$

Then 2-25 becomes, if $I = \{1,2,..,A\}$:

$$y^T B(x) y + \sum_{\alpha \in I} \xi_\alpha(x,y) \geq \varepsilon |y|^2 \frac{1}{1+|x|^\delta}. \tag{7-35}$$

If $\xi_\alpha(x,y) = 0$ for all x,y we can delete this value α from I without altering 7-35. If $\xi_\alpha(x,y) > 0$ for some pair (x,y), then $f_\alpha(z) \leq \xi_\alpha(x,y)^{-1} \rho_\alpha(z) y^T C_\alpha(x,z) y$ for all $z \in E_\alpha$. But (A'-2) and (SC) imply that $|C_\alpha(x,.)| \leq \theta'(1+\eta_\alpha)^2$ where $\theta' > 0$ and $\eta_\alpha \in \cap_{2 \leq p < \infty} L^p(E_\alpha, G_\alpha)$, while

$\rho_\alpha \in \cap_{1 \le p < \infty} L^p(E_\alpha, G_\alpha)$. Thus clearly $f_\alpha \in L^1(E_\alpha, G_\alpha)$. In particular $G_\alpha(f_\alpha > 1) < \infty$. Thus Lemma 2-34 implies that $f_\alpha \wedge 1$ is still (ζ,θ)-broad. Moreover, replacing f_α by $f_\alpha \wedge 1$ increases $\xi_\alpha(x,y)$ and so does not alter 7-35. That is to say, we may assume $f_\alpha \le 1$. Set

$$\gamma_\alpha = \int_{E_\alpha} f_\alpha(z)dz \quad , \qquad \gamma = 1 \wedge \frac{\varepsilon}{A+1} \wedge \min_{\alpha \in I} \gamma_\alpha .$$

We have $\gamma_\alpha \le \gamma$, and if $\gamma_\alpha < \gamma$ we may replace f_α by $f_\alpha \gamma/\gamma_\alpha$ without altering neither 7-35 nor the (ζ,θ)-broadness of f_α. In other words, we may and will assume that for all $\alpha \in I$:

$$0 \le f_\alpha \le 1, \quad \int_{E_\alpha} f_\alpha(z)dz = \gamma \le 1 \wedge \frac{\varepsilon}{A+1}. \tag{7-36}$$

<u>Step 2</u>. Until further notice, we drop the superscript "x". Let $\sigma \in S_{d-1}$ be given, and set

$$V = \{(\omega,t) : \sigma^T \nabla X_{t-}^{-1}(\omega) B(X_{t-}(\omega)) \nabla X_{t-}^{-1,T}(\omega)\sigma$$
$$\ge \gamma^2 |\nabla X_{t-}^{-1,T}(\omega)\sigma|^2 / (1+|X_{t-}(\omega)|^\delta)\},$$

$$V'_\alpha = \{(\omega,t) : \sigma^T \nabla X_{t-}^{-1}(\omega) C_\alpha(X_{t-}(\omega),z) \nabla X_{t-}^{-1,T}(\omega)\sigma \rho_\alpha(z)$$
$$\ge \gamma f_\alpha(z) |\nabla X_{t-}^{-1,T}(\omega)\sigma|^2/(1+|X_{t-}(\omega)|^\delta) \quad \text{for all } z \in E_\alpha\}$$

if $\alpha \in I$, and $V'_\alpha = \emptyset$ if $\alpha \notin I$,

$$V_\alpha = V'_\alpha \setminus (V \cup V'_1 \cup \ldots \cup V'_{\alpha-1}).$$

Since $\gamma \le 1 \wedge \frac{\varepsilon}{1+A}$ we deduce from 7-35 that

$$(V,(V_\alpha)_{\alpha \in I}) \text{ is a predictable partition of of } \Omega \times [0,T]. \tag{7-37}$$

<u>Step 3</u>. Let $H_t = \delta' |\nabla X_t|^{\delta''} (1+|X_t|^{\delta''})^{-1} (1+|\nabla X_t|^{\delta''})^{-1}$.

Since all the integrands of 7-32 are nonnegative matrices, we deduce from 7-33 that

$$\sigma^T S\sigma \geq H_-\sigma^T \nabla X_-^{-1} B(X_-)\nabla X_-^{-1,T}\sigma\, 1_V * t$$
$$+ \sum_{\alpha\in I} H_-\sigma^T \nabla X_-^{-1} C_\alpha(X_-)\nabla X_-^{-1,T}\sigma\rho_\alpha 1_{V_\alpha} * \mu_\alpha$$
$$\geq \frac{H_-}{1+|X_-|^\delta}\gamma^2 |\nabla X_-^{-1,T}\sigma|^2\, 1_V * t$$
$$+ \sum_{\alpha\in I} \frac{H_-}{1+|X_-|^\delta}\gamma|\nabla X_-^{-1,T}\sigma|^2\, f_\alpha 1_{V_\alpha} * \mu_\alpha.$$

Call $\|\nabla X_{t-}^T\|$ the operator norm of ∇X_{t-}^T: then $|\nabla X_{t-}^{-1,T}\sigma|^2 \geq \|\nabla X_{t-}^T\|^{-2}$ (because $|\sigma|=1$). Thus if

$$F_t = \sup_{s\leq t} \|\nabla X_s^T\|^2\, \delta'^{-1}(1+|X_s|^\delta)(1+|X_s|^{\delta''})(1+|\nabla X_s|^{\delta''})|\nabla X_s|^{-\delta''}, \tag{7-38}$$

then

$$F_t\sigma^T S_t\sigma \geq \gamma^2 \int_0^t 1_V(s)ds + \sum_\alpha \gamma f_\alpha 1_{V_\alpha} * \mu_\alpha.$$

Since $f_\alpha\in L^1(E_\alpha,G_\alpha)$ the stochastic integral $Z^\alpha = f_\alpha 1_{V_\alpha} * \tilde\mu_\alpha$ exists and

$$Z^\alpha = (f_\alpha 1_{V_\alpha})*\mu_\alpha - (f_\alpha 1_{V_\alpha})*\nu_\alpha = (f_\alpha 1_{V_\alpha})*\mu_\alpha - \gamma 1_{V_\alpha} * t$$

(recall 7-36). Thus with $Z=\sum_\alpha Z^\alpha$, and using 7-37, we get

$$F_t\sigma^T S_t\sigma \geq \gamma(\gamma t + Z_t). \tag{7-39}$$

Step 4. We shall now compute $E(Y_t^\phi)$, where $Y^\phi = e^{-\phi Z}$ and $\phi>0$. Ito's formula gives (as in 7-11):

$$Y^\phi = 1 - \sum_{\alpha\in I} \phi Y_-^\phi f_\alpha 1_{V_\alpha} * \tilde\mu_\alpha + \sum_{\alpha\in I} Y_-^\phi(e^{-\phi f_\alpha} - 1 + \phi f_\alpha) 1_{V_\alpha} * \mu_\alpha$$
$$= 1 + \sum_{\alpha\in I} [\phi\gamma Y_-^\phi 1_{V_\alpha} * t + Y_-^\phi(e^{-\phi f_\alpha} - 1) 1_{V_\alpha} * \mu_\alpha]$$

(use again 7-36). Since $0\leq 1-e^{-\phi f_\alpha} \leq \phi f_\alpha \in L^1(E_\alpha,G_\alpha)$, we can define the nonnegative numbers:

$$\eta_\alpha^\phi = \phi\gamma - \int_{E_\alpha} (1-e^{-\phi f_\alpha(z)})dz \tag{7-40}$$

and rewrite

$$Y^\phi = 1 + \sum_{\alpha\in I} \{Y_-^\phi(e^{-\phi f_\alpha}-1)1_{V_\alpha} * \tilde{\mu}_\alpha + \eta_\alpha^\phi Y_-^\phi 1_{V_\alpha} * t\}. \tag{7-41}$$

Using again $0\leq 1-e^{-\phi f_\alpha} \leq \phi f_\alpha$, we can apply Theorem 5-10 to Equation 7-41 in order to obtain that $|Y^\phi|_T^* \in \cap_{q<\infty} L^q$, and since $f_\alpha \in \cap_{1\leq q<\infty} L^q(E_\alpha, G_\alpha)$, Lemma 5-1 yields that the local martingale $Y_-^\phi(\exp(-\phi f_\alpha)-1)1_{V_\alpha} * \tilde{\mu}_\alpha$ is in fact a martingale. Then, taking expectations in 7-41 yields

$$E(Y_t^\phi) = 1 + \sum_{\alpha\in I} \eta_\alpha^\phi E(\int_0^t Y_s^\phi 1_{V_\alpha}(s)ds)$$
$$1 + (\max_{\alpha\in I} \eta_\alpha^\phi)\int_0^t E(Y_s^\phi)ds,$$

and Gronwall's Lemma gives

$$E(e^{-\phi Z_t}) = E(Y_t^\phi) \leq \exp(t \max_{\alpha\in I} \eta_\alpha^\phi). \tag{7-42}$$

<u>Step 5</u>. For $\sigma\in S_{d-1}$, $u>0$, $v>0$, $t>0$, set

$$h_{u,v}(\sigma,t) = \int_0^\infty s^{u-1}\, ds\, \{E[\exp -s\, F_t(\sigma^T S_t \sigma)]\}^v. \tag{7-43}$$

Using 7-39, 7-40 and 7-42, we obtain

$$E\{\exp -\frac{\phi}{\gamma} F_t(\sigma^T S_t \sigma)\} \leq \exp(-t\gamma\phi + t \max_{\alpha\in I} \eta_\alpha^\phi)$$
$$\leq \max_{\alpha\in I} \exp[-t \int_{E_\alpha}(1-e^{-\phi f_\alpha(z)})dz].$$

Taking $\phi=\gamma s$ above, we deduce (with notation 2-21):

$$h_{u,v}(\sigma,t) \leq \int_0^\infty s^{u-1}\, ds \max_{\alpha\in I} \exp[-vt \int_{E_\alpha}(1-e^{-\gamma s f_\alpha(z)})dz]$$
$$\leq \gamma^{-u} \sum_{\alpha\in I} \gamma_\alpha(u,vt,f_\alpha). \tag{7-44}$$

Step 6. Let $p>0$, $n\in\mathbb{N}^*$. 7-29 gives

$$\det(F_t S_t)^{-np} \leq \frac{1}{2^n\Gamma(p)^{nd}} \int_0^\infty\cdots\int_0^\infty (s_1\cdots s_n)^{pd-1}\, ds_1\cdots ds_n \int_{S_{d-1}}\cdots\int_{S_{d-1}} \prod_{i=1}^n e^{-s_i F_t \sigma_i^T S_t \sigma_i}\, d\sigma_1\cdots d\sigma_n.$$

Since

$$E\Big(\prod_{i=1}^n |W_i|\Big) \leq \prod_{i=1}^n E(|W_i|^n)^{1/n}$$

for all r.v. W_i, we obtain from 7-43 and 7-44 by taking expectations:

$$E(\det(F_t S_t)^{-np}) \leq \frac{1}{2^n\Gamma(p)^{nd}} \int_0^\infty\cdots\int_0^\infty (s_1\cdots s_n)^{pd-1} ds_1\cdots ds_n \int_{S_{d-1}}\cdots\int_{S_{d-1}} \prod_{i=1}^n E(e^{-ns_i F_t\sigma_i^T S_t \sigma_i})^{1/n} d\sigma_1\cdots d\sigma_n$$

$$\leq \frac{1}{2^n\Gamma(p)^{nd}}\frac{1}{n^{npd}} \Big\{\int_{S_{d-1}} h_{pd,1/n}(\sigma,t)\ d\sigma\Big\}^n$$

$$\leq \delta_{p,n,\gamma,d}\Big\{\sum_{\alpha\in I} \gamma_\alpha(pd,\tfrac{t}{n},f_\alpha)\Big\}^n \tag{7-45}$$

where $\delta_{p,n,\gamma,d}$ is a constant depending only on p,n,γ,d.

Step 7. Now we make the dependence upon the starting point x explicit. Let p,t,n with $pd<\zeta$, $\frac{t}{n}\geq\theta$, and $n\in\mathbb{N}^*$. Let $\varepsilon>0$ be such that $(p+\frac{\varepsilon}{n})d<\zeta$. Since each f_α is (ζ,θ)-broad, 7-45 gives a constant δ (not depending on x) such that for all $x\in\mathbb{R}^d$, and if $p'=p+\frac{\varepsilon}{n}$,

$$E(\det(F_t^x\ S_t^x)^{-np'}) \leq \delta. \tag{7-46}$$

We know that $x \rightsquigarrow \|(X^x)_T^*\|_{L^q}$ and $x\rightsquigarrow \|(\nabla X^x)_T^*\|_{L^q}$ and $x\rightsquigarrow\|((\nabla X^x)^{-1})_T^*\|_{L^q}$ (by 7-27) are locally bounded for all $q<\infty$, and so is $x\rightsquigarrow\|F_t^x\|_{L^q}$ because of 7-38. We also have $\det(S_t^x)^{-1} = (F_t^x)^d \det(F_t^x S_t^x)^{-1}$, hence 7-34 follows from 7-46 and the fact that $p'>p$.

7-47 <u>Proof of Theorems 2-27, 2-28, 2-29 under (A'-r)</u>. We will apply Theorem 6-48. We assume $(SB-(\zeta,\theta))$ with some ρ_α like in 2-23, and (SC) and at least (A'-3), and according to 6-49 we put $q_t^x(i)=E(|\det(DX_t^x)|^{-i})$.

The choice of u and v_α is as in Lemma 7-14. We firstly consider h,h'_α,h'' as in 7-18. If ζ is the constant showing in (SC), it is obviously possible to choose h'' so that

$$h''(y) = \begin{cases} 1 & \text{if } |\det(y)| \geq \zeta \\ \dfrac{\det(y)^2}{1+|y|^{4d}} & \text{if } |\det(y)| \leq \frac{\zeta}{2}. \end{cases} \tag{7-48}$$

Then f,g_α are given by 7-19. Because of 7-2 and (SC), we have $|\det(I+\Delta K^x)|\geq\zeta$ identically, so $T_1^x=\infty$ in 6-35 and 6-37 yields

$$DX_t^x = \nabla X_t^x \int_0^t (\nabla X_{s-}^x)^{-1} \; (I+\Delta K_s^x) \; dH_s^x.$$

Then due to 7-20, we get

$$DX_t^x = \nabla X_t^x \int_0^t (\nabla X_{s-}^x)^{-1} \; d\hat{R}_s^x \; (\nabla X_{s-}^x)^{-1,T},$$

where $\hat{R}^x$ is given by 7-21. Now, if we compare 7-21 and 7-32, we get

$$DX_t^x = \nabla X_t^x \; S_t^x \;, \tag{7-49}$$

provided

$$k(x,y) = h(b(x))h''(y)$$

$$k'_\alpha(x,y,z) = h'_\alpha(D_z c_\alpha(x,z))h''(I+D_x c_\alpha(x,z))h''(y).$$

In virtue of 7-18 and 7-48 and (A'-1) and (SC), one easily checks that 7-33 is met for some constants δ',

$\delta''>0$. Hence by Lemma 7-31, if $n\in\mathbb{N}^*$ and $p\in(0,\infty)$ satisfy $pd<\zeta$ and $\frac{t}{n}\geq\theta$, there is a constant $\delta_{A,t}$ for each $t>0$ and each bounded subset A of $\mathbb{R}^d$ with

$$E(\det(S_s^x)^{-np}) \leq \delta_{A,t} \quad \text{for all } x\in A, s\geq t \tag{7-50}$$

(recall that $s \rightsquigarrow \det(S_s^x)$ increases). By 7-27, $x \rightsquigarrow \sup_{s\leq T} E(|\det(\nabla X_s^x)|^{-q})$ is locally bounded for all $q<\infty$. Moreover $\det(DX_s^x) = \det(\nabla X_s^x)\det(S_s^x)$ by 7-49. Therefore 7-50 yields that

$$E(|\det(DX_s^x)|^{-np'}) \leq \delta'_{A,t} \quad \text{for all } x\in A, s\geq t,$$

provided $p'<p$, and where $\delta'_{A,t}$ is a constant depending on $\delta_{A,t}$ and on p'. In other words,

$$\left.\begin{array}{l} x \rightsquigarrow \sup_{s\geq t} q_s^x(i) \text{ is locally bounded if} \\ \quad i=np', \text{ where } n\in\mathbb{N}^*, \ \frac{t}{n}\geq\theta \text{ and } p'd<\zeta. \end{array}\right\} \tag{7-51}$$

Observe that the conditions in 7-51 are equivalent to saying that $t\geq\theta$ and $\zeta>d\, i/[\frac{t}{\theta}]$. Hence parts (b), (c), (d) of Theorem 6-48 give Theorems 2-27, 2-28, 2-29 respectively (in the setting of 2-29, 6-50 should of course be replaced by 2-30).

CHAPTER IV

MALLIAVIN'S APPROACH

Section 8: MALLIAVIN OPERATORS

This last chapter expounds Malliavin's (or Stroock's, or Shigekawa's) approach for finding a (smooth) density for the random variables X_t^x in 1-4. Sections 8 and 9, which are independent of what precedes (except for the notation of §6-a and the definition 4-4 of an integration-by-parts setting) are concerned with general facts. Section 10 applies to stochastic differential equations. The (very short) Section 11 is a replica of Section 7, in which the main results of Section 2 are re-proved. Finally, Section 12 contains some heuristic remarks about Malliavin operators and their connections with Bismut's approach.

As a matter of fact, the present Section 8 is concerned with the same abstract setting as Section 4: it explains how Malliavin's ideas can be set for constructing an integration-by-parts setting which works simultaneously for a "large" collection of random variables. Stroock also has introduced "abstract" Malliavin operators in [26], following a slightly different approach.

§8-a. DEFINITION OF MALLIAVIN OPERATORS

Throughout we fix a probability space $(\Omega,\underline{\underline{F}},P)$. Recall that $C_p^2(R^n)$ denotes the set of all functions on $\mathbb{R}^n$ of class C^2, whose derivatives have at most polynomial growth.

8-1 <u>DEFINITION</u>: A *Malliavin operator* $(L,\mathcal{R})$ is a linear operator L on a domain $\mathcal{R}\subset\cap_{p<\infty}L^p$, taking values in $\cap_{p<\infty}L^p$, and such that:

(i) $\mathcal{R}$ is stable under C_p^2: i.e., if $\Phi\in\mathcal{R}$, $F\in C_p^2(R^n)$, then $F(\Phi)\in\mathcal{R}$.

(ii) $\underline{\underline{F}}$ is the σ-field generated by all functions in $\mathcal{R}$.

(iii) L is self-adjoint in L^2, i.e. $E(\Phi L\Psi)=E(\Psi L\Phi)$ for all $\Phi,\Psi\in\mathcal{R}$.

(iv) $L(\Phi^2)\geq 2\Phi L\Phi$ for $\Phi\in\mathcal{R}$; this amounts to saying that the symmetric bilinear operator Γ associated with L by

$$\Gamma(\Phi,\Psi) = L(\Phi\Psi) - \Phi L\Psi - \Psi L\Phi \tag{8-2}$$

on $\mathcal{R}\times\mathcal{R}$ is nonnegative.

(v) If $\Phi=(\Phi^1,\ldots,\Phi^n)\in\mathcal{R}^n$ and $F\in C_p^2(R^n)$, then

$$L(F\circ\Phi) = \sum_{i=1}^{n}\frac{\partial}{\partial x_i}F(\Phi)\,L\Phi^i + \frac{1}{2}\sum_{i,j=1}^{n}\frac{\partial^2}{\partial x_i\partial x_j}F(\Phi)\Gamma(\Phi^i,\Phi^j). \tag{8-3}$$

Here are some easy consequences of this definition. Firstly, one notices that 8-2 and 8-3 are compatible. Secondly, an easy computation based upon 8-2 and 8-3 shows that if $\Phi=(\Phi^1,\ldots,\Phi^k)\in\mathcal{R}^k$, $\Psi\in\mathcal{R}$, $F\in C_p^2(R^k)$, then

$$\Gamma(F(\Phi),\Psi) = \sum_{i=1}^{k}\frac{\partial}{\partial x_i}F(\Phi)\;\Gamma(\Phi^i,\Psi). \tag{8-4}$$

We also have $\Gamma(\Phi,\Psi)\in\cap_{p<\infty}L^p$ if $\Phi,\Psi\in\mathcal{R}$. From (v), $L1=0$; hence (iii) and (iv) yield for $\Phi,\Psi\in\mathcal{R}$:

$$\left.\begin{aligned} &E(L\Phi) = 0 \\ &E(\Phi L\Psi) = E(\Psi L\Phi) = -\frac{1}{2}E(\Gamma(\Phi,\Psi)). \end{aligned}\right\} \quad (8\text{-}5)$$

The nonnegativeness of Γ implies

$$|\Gamma(\Phi,\Psi)| \le \Gamma(\Phi,\Phi)^{1/2}\ \Gamma(\Psi,\Psi)^{1/2}, \quad (8\text{-}6)$$

which in turn yields

$$\Gamma(\Phi+\Psi,\Phi+\Psi)^{1/2} \le \Gamma(\Phi,\Phi)^{1/2} + \Gamma(\Psi,\Psi)^{1/2}; \quad (8\text{-}7)$$

We also deduce from 8-5 that

$$\|\Gamma(\Phi,\Phi)\|_{L^1} \le 2\ \|\Phi\|_{L^2}\ \|L\Phi\|_{L^2}. \quad (8\text{-}8)$$

The following fundamental lemma is well known for self-adjoint operators:

8-9 <u>LEMMA</u>: *Let* $\{\Phi_n\}\subset\mathcal{R}$ *with* $\Phi_n \to 0$ *and* $L\Phi_n \to \theta$ *in* L^2. *Then* $\theta=0$ *a.s., and* $\Gamma(\Phi_n,\Phi_n) \to 0$ *in* L^1.

<u>Proof</u>. For all $\Psi\in\mathcal{R}$ we have $E(\Psi L\Phi_n)=E(\Phi_n L\Psi)$; going to the limit as $n\uparrow\infty$ yields $E(\Psi\theta)=0$, and 8-1-(ii) implies $\theta=0$ a.s. We also have $E(\Gamma(\Phi_n,\Phi_n))=-2E(\Phi_n L\Phi_n)$, from which we deduce the second claim.

Finally, the Malliavin operator $(L,\mathcal{R})$ pertains to the construction 4-4 as such:

8-10 PROPOSITION: *If* $\Phi=(\Phi^1,\ldots,\Phi^k)\in\mathcal{R}^k$, *the following set* $(\sigma_\Phi,\gamma_\Phi,\mathcal{H}_\Phi,\gamma_\Phi)$ *is an integration-by-parts setting for* Φ:

$$\left.\begin{aligned} &\sigma_\Phi^{ij} = \Gamma(\Phi^i,\Phi^j), \quad \gamma_\Phi^j = -2L\Phi^j \\ &\mathcal{H}_\Phi = \mathcal{R},\ \delta^j(\Psi) = -\Gamma(\Phi^j,\Psi) \quad \text{for } \Psi\in\mathcal{R}. \end{aligned}\right\} \quad (8\text{-}11)$$

<u>Proof</u>. These terms clearly meet conditions (i)-(iv) of

4-4 (4-5 follows from 8-4). Let $f\in C_p^2(R^k)$. Then 4-6 follows from

$$E(\Psi \sum_{i=1}^{k} \frac{\partial}{\partial x_i} f(\Phi)\sigma_\Phi^{ij}) = E[\Psi\Gamma(f\circ\Phi,\Phi^j)] \qquad \text{(by 8-4)}$$

$$= E[\Psi L(\Phi^j f(\Phi)) - \Psi\Phi^j L(f\circ\Phi) - \Psi f(\Phi)L\Phi^j] \qquad \text{(by 8-2)}$$

$$= E[\Phi^j f(\Phi)L\Psi - f(\Phi)L(\Psi\Phi^j) - \Psi f(\Phi)L\Phi^j] \qquad \text{(by 8-5)}$$

$$= E[f(\Phi)\{\Phi^j L\Psi - \Gamma(\Psi,\Phi^j) - \Psi L\Phi^j - \Phi^j L\Psi - \Psi L\Phi^j\}] \qquad \text{(by 8-2)}$$

$$= E[f(\Phi)\{\delta_\Phi^j(\Psi) + \Psi\gamma_\Phi^j\}].$$

§8-b. EXTENSION OF MALLIAVIN OPERATORS

Despite (i) and (ii) of 8-1, the a-priori domain $\mathcal{R}$ might be too small to contain the variables of interest. This is why we now proceed to extending $\mathcal{R}$, essentially following Stroock [26].

8-12 DEFINITION: We call H_2 the set of all $\Phi\in L^2$ for which there is a sequence $\{\Phi_n\}\subset\mathcal{R}$ going to Φ in L^2, such that $\{L\Phi_n\}$ is Cauchy in L^2; we then denote by $L\Phi$ the L^2-limit of $L\Phi_n$ (due to 8-9, this limit does not depend on the particular sequence $\{\Phi_n\}$). For $\Phi\in H_2$ we set

$$\|\Phi\|'_{H_2} = \|\Phi\|_{L^2} + \|L\Phi\|_{L^2}$$

(the graph-norm of (L,H_2)).

8-13 PROPOSITION: *a)* $(H_2,\|.\|'_{H_2})$ *is a Banach space, and is the closure of* $\mathcal{R}$ *for the norm* $\|.\|'_{H_2}$.

b) Γ *can be extended uniquely as a continuous bilinear symmetric operator:* $H_2\times H_2 \to L^1$. *Moreover, this extension is nonnegative.*

c) (L,H_2) *is a self-adjoint (closed) operator on* L^2, *and 8-5, 8-6, 8-7, 8-8 hold for all* $\Phi,\Psi\in H_2$.

Proof. (a) is trivial. To prove (b) and (c), let $\{\Phi_n\}$, $\{\Psi_n\}\subset\mathcal{R}$ going to Φ and Ψ for $\|.\|'_{H_2}$. From 8-6 and 8-8,

$$E(|\Gamma(\Phi_n,\Psi_n)-\Gamma(\Phi_m,\Psi_m)|)$$
$$\leq E(|\Gamma(\Phi_n-\Phi_m,\Psi_n)|) + E(|\Gamma(\Phi_m,\Psi_n-\Psi_m)|)$$
$$\leq 2\{\|\Phi_n-\Phi_m\|_{L^2}\|L\Phi_n-L\Phi_m\|_{L^2}\|\Psi_n\|_{L^2}\|L\Psi_n\|_{L^2}\}^{1/2}$$
$$+ 2\{\|\Phi_m\|_{L^2}\|L\Phi_m\|_{L^2}\|\Psi_n-\Psi_m\|_{L^2}\|L\Psi_n-L\Psi_m\|_{L^2}\}^{1/2},$$

which goes to 0 as $n,m\uparrow\infty$. Hence $\Gamma(\Phi_n,\Psi_n)$ goes to a limit, say $\Gamma(\Phi,\Psi)$, in L^1; moreover if $\{\Phi'_n\},\{\Psi'_n\}\subset\mathcal{R}$ also converge to Φ and Ψ for $\|.\|'_{H_2}$, the same argument applied to $\Gamma(\Phi_n,\Psi_n)-\Gamma(\Phi'_n,\Psi'_n)$ shows that $\Gamma(\Phi'_n,\Psi'_n)\to\Gamma(\Phi,\Psi)$ as well. Hence Γ is a bilinear symmetric nonnegative operator: $H_2\times H_2\to L^1$, which extends 8-2. Therefore we have 8-6 and 8-7, and also 8-5 and 8-8 by passing to the limit.

Finally if $\{\Phi_n\},\{\Psi_n\}\subset H_2$ converge to Φ and Ψ for $\|.\|'_{H_2}$, again the same argument shows that $\Gamma(\Phi_n,\Psi_n)\to\Gamma(\Phi,\Psi)$ in L^1. So we have (b), and (c) readily follows from all what precedes.

Of course the domain H_2 is not in $\cap_{p<\infty}L^p$, and it is not stable under C^2_p, so (L,H_2) is not a Malliavin operator. So we presently restrict the domain of (L,H_2). Firstly, set

$$\|\Phi\|_{H_q} = \|\Phi\|_{L^q} + \|L\Phi\|_{L^q} + \|\Gamma(\Phi,\Phi)^{1/2}\|_{L^q} \qquad (8\text{-}14)$$

for $q\geq 2$ and $\Phi\in H_2$, with the convention $\|\Phi\|_{H_q}=\infty$ if either Φ or $L\Phi$ or $\Gamma(\Phi,\Phi)^{1/2}$ does not belong to L^q. Then

$\|\Phi\|_{H_q} < \infty$ when $\Phi \in \mathcal{R}$ and, due to 8-7, the set $\{\Phi \in H_2 :$ $\|\Phi\|_{H_q} < \infty\}$ is a linear space on which $\|.\|_{H_q}$ is a norm. From 8-8 we deduce

$$\|\Phi\|'_{H_2} \leq \|\Phi\|_{H_2} \leq 2\|\Phi\|'_{H_2}. \tag{8-15}$$

8-16 DEFINITION: Let $q \in [2,\infty)$. Then H_q denotes the closure of $\mathcal{R}$ under $\|.\|_{H_q}$ (by 8-15 this definition coincides with 8-12 when $q=2$; note that $H_q \supset H_{q'}$ for $q'>q$). Set also $H_\infty = \cap_{2 \leq q < \infty} H_q$.

8-17 PROPOSITION: *Let* $q \in [2,\infty)$.

a) $(H_q, \|.\|_{H_q})$ *is a Banach space.*

b) Let $\{\Phi_n\} \subset H_q$ *converge to* Φ *in* L^q. *Then* $\|\Phi_n - \Phi\|_{H_q} \to 0$ (which in particular implies that $\Phi \in H_q$) *if and only if* $\{L\Phi_n\}$ *is Cauchy in* L^q *and if the sequence* $\{\Gamma(\Phi_n,\Phi_n)^{q/2}\}$ *is UI*(= Uniformly Integrable). *In this case, we also have* $\Gamma(\Phi_n,\Phi_n) \to \Gamma(\Phi,\Phi)$ *in* $L^{q/2}$.

Proof. For $q=2$ the claims follow from 8-13 and 8-15. For $q>2$ we begin with (b). Let $\{\Phi_n\} \subset H_q$ go to 0 in L^q.

(i) Assume first that $\|\Phi_n - \Phi\|_{H_q} \to 0$. Then $L\Phi_n \to L\Phi$ in L^q. From 8-7 we deduce

$$\Gamma(\Phi_n,\Phi_n)^{q/2} \leq 2^{q-1}[\Gamma(\Phi_n-\Phi,\Phi_n-\Phi)^{q/2} + \Gamma(\Phi,\Phi)^{q/2}],$$

hence the sequence $\{\Gamma(\Phi_n,\Phi_n)^{q/2}\}$ is UI. Moreover, $\Gamma(\Phi_n,\Phi_n) \to \Gamma(\Phi,\Phi)$ in L^1 by 8-13, so this convergence also holds in $L^{q/2}$.

(ii) Conversely, assume that $\{L\Phi_n\}$ is Cauchy in L^q and that $\{\Gamma(\Phi_n,\Phi_n)^{q/2}\}$ is UI. From 8-12 we deduce that $\Phi \in H_2$, and $L\Phi_n \to L\Phi$ in L^2, and thus $L\Phi_n \to L\Phi$ in L^q as well. Moreover $\Gamma(\Phi_n,\Phi_n) \to \Gamma(\Phi,\Phi)$ in L^1 by 8-13, thus $\Gamma(\Phi_n,\Phi_n) \to \Gamma(\Phi,\Phi)$ in $L^{q/2}$ by the UI property. Since

$$\Gamma(\Phi_n-\Phi,\Phi_n-\Phi)^{q/2} \leq 2^{q-1}[\Gamma(\Phi_n,\Phi_n)^{q/2} + \Gamma(\Phi,\Phi)^{q/2}]$$

and $\Gamma(\Phi_n-\Phi,\Phi_n-\Phi) \to 0$ in L^1 by 8-13, we obtain that $\Gamma(\Phi_n-\Phi,\Phi_n-\Phi) \to 0$ in $L^{q/2}$, and $\|\Phi_n-\Phi\|_{H_q} \to 0$ follows.

It remains to prove (a). Let $\{\Phi_n\}\subset H_q$ be Cauchy for $\|.\|_{H_q}$. Then there exists $\Phi\in L^q$ such that $\Phi_n \to \Phi$ in L^q, and $\{L\Phi_n\}$ is Cauchy in L^q. Finally the family $\{\Gamma(\Phi_n-\Phi_m,\Phi_n-\Phi_m)^{q/2}\}_{n,m\geq 1}$ is UI, while

$$\Gamma(\Phi_n,\Phi_n)^{q/2} \leq 2^{q-1}[\Gamma(\Phi_m,\Phi_m)^{q/2}+\Gamma(\Phi_n-\Phi_m,\Phi_n-\Phi_m)^{q/2}].$$

Hence $\{\Gamma(\Phi_n,\Phi_n)^{q/2}\}$ is UI. Thus (b) implies that $\Phi\in H_q$ and $\|\Phi_n-\Phi\|_{H_q} \to 0$.

8-18 <u>THEOREM</u>: *The operator* (L,H_∞) *is a Malliavin operator* (hence, the conclusions of Proposition 8-10 hold with H_∞ instead of $\mathcal{R}$).

<u>Proof</u>. (L,H_∞) satisfies 8-1-(ii) (because $\mathcal{R}\subset H_\infty$) and 8-1-(iii) (by 8-13-c), while the restriction of Γ to $H_\infty\times H_\infty$ is nonnegative (by 8-13-b). We also have $H_\infty\subset\cap_{p<\infty}L^p$, and $L\Phi\in\cap_{p<\infty}L^p$ if $\Phi\in H_\infty$. It remains to prove that if $\Phi=(\Phi^1,\ldots,\Phi^k)\in(H_\infty)^k$, $F\in C_p^2(R^k)$, then $\Psi=F(\Phi)$ belongs to H_∞ and $L\Psi$ is given by 8-3 (since 8-2 is a particular case of 8-3, we will thus have 8-1-(iv) as well).

Let $q\in[2,\infty)$. There is a sequence $\{\Phi_n=(\Phi_n^1,\ldots,\Phi_n^k)\}\subset\mathcal{R}^k$ such that $\|\Phi_n^i-\Phi^i\|_{H_q} \to 0$ for all $i\leq k$. Set $\Psi_n=F(\Phi_n)$. Since $\Phi_n\in\mathcal{R}^k$, 8-3 and 8-4 give

$$L\Psi_n = \sum_{i=1}^{k} \frac{\partial}{\partial x_i}F(\Phi_n)L\Phi_n^i + \frac{1}{2}\sum_{i,j=1}^{k}\frac{\partial^2}{\partial x_i\partial x_j}F(\Phi_n)\Gamma(\Phi_n^i,\Phi_n^j)$$

$$\Gamma(\Psi_n,\Psi_n)= \sum_{i,j=1}^{k} \frac{\partial}{\partial x_i}F(\Phi_n)\frac{\partial}{\partial x_j}F(\Phi_n)\Gamma(\Phi_n^i,\Phi_n^j). \tag{8-19}$$

Let $\zeta,\theta\geq 0$ be such that $|F(x)|,|D_xF(x)|,\ |D^2_{x^2}F(x)|\leq \zeta(1+|x|^\theta)$. Since $\Phi_n\to\Phi$ in L^q, the r.v. $F(\Phi_n)$, $\partial F/\partial x_i(\Phi_n)$, $\partial^2F/\partial x_i\partial x_j(\Phi_n)$ obviously converge to $F(\Phi)$, $\partial F/\partial x_i(\Phi)$, $\partial^2F/\partial x_i\partial x_j(\Phi)$ in $L^{q/\theta}$. We also know that $L\Phi^i_n\to L\Phi^i$ in L^q and $\Gamma(\Phi^i_n,\Phi^j_n)\to\Gamma(\Phi^i,\Phi^j)$ in $L^{q/2}$. Then, passing to the limit in 8-19 as $n\uparrow\infty$, we see that $L\Psi_n$ goes in $L^{q/(2+\theta)}$ to the r.v. $L\Psi$ defined by 8-3, and that $\Gamma(\Psi_n,\Psi_n)$ also goes in $L^{q/(2+\theta)}$ to

$$\Gamma(\Psi,\Psi)\ =\ \sum_{i,j=1}^{k}\frac{\partial}{\partial x_i}F(\Phi)\ \frac{\partial}{\partial x_j}F(\Phi)\ \Gamma(\Phi^i,\Phi^j),$$

while we have already seen that $\Psi_n\to\Psi$ is $L^{q/\theta}$. But $\Psi_n\in\mathcal{R}$, so it follows that Ψ belongs to $H_{q/(2+\theta)}$, provided $q/(2+\theta)\geq 2$, and it also follows that $L\Psi$ is given by 8-3. Since q is arbitrarily large we deduce $\Psi\in H_\infty$.

§8-c. MALLIAVIN OPERATORS ON A DIRECT PRODUCT

This subsection is mainly technical. It will be used essentially to prove the "adaptedness" of Malliavin operators on Poisson-Wiener space.

We start with two probability spaces $(\Omega',\underline{\underline{F}}',P')$ and $(\Omega'',\underline{\underline{F}}'',P'')$. On each one there is a Malliavin operator $(L',\mathcal{R}')$ (resp. $(L'',\mathcal{R}'')$), with which are associated Γ', H'_∞, H'_q (resp. Γ'', H''_∞, H''_q) as in §8-b. Set

$$(\Omega,\underline{\underline{F}},P)\ =\ (\Omega',\underline{\underline{F}}',P')\otimes(\Omega'',\underline{\underline{F}}'',P''). \tag{8-20}$$

If Φ' (resp. Φ'') is a function on Ω' (resp. Ω'') we denote by the same symbol its natural extension to Ω. If Φ is a function on Ω, we denote by $\Phi'_{\omega''}(\omega')=\Phi(\omega',\omega'')$ and $\Phi''_{\omega'}(\omega'')=\Phi(\omega',\omega'')$ its "sections". We also write E' and E'' for the expectations with respect to P' and P''.

8-21 DEFINITION: The *direct product Malliavin operator* $(L,\mathcal{R})$ is defined by:

(i) $\mathcal{R}$ is the set of functions of the form $\Phi=F(\Phi'_1,\ldots,\Phi'_n;\Phi''_1,\ldots,\Phi''_d)$, where $\Phi'_i\in\mathcal{R}'$, $\Phi''_i\in\mathcal{R}''$, $F\in C^2_p(R^n\times R^d)$, $n,d\in\mathbb{N}^*$.

(ii) If Φ is as above, we obviously have $\Phi'_{\omega''}\in\mathcal{R}'$ and $\Phi''_{\omega'}\in\mathcal{R}''$, and we set

$$L\Phi(\omega',\omega'') = L'\Phi'_{\omega''}(\omega') + L''\Phi''_{\omega'}(\omega''). \tag{8-22}$$

8-23 PROPOSITION: *The direct product Malliavin operator* $(L,\mathcal{R})$ *is a Malliavin operator on* $(\Omega,\underline{\underline{F}},P)$, *and the associated bilinear symmetric operator* Γ *satisfies for* $\Phi,\Psi\in\mathcal{R}$:

$$\Gamma(\Phi,\Psi)(\omega',\omega'') = \Gamma'(\Phi'_{\omega''},\Psi'_{\omega''})(\omega') + \Gamma''(\Phi''_{\omega'},\Psi''_{\omega'})(\omega''). \tag{8-24}$$

Proof. That $\mathcal{R}\subset\cap_{p<\infty}L^p$ and that $\mathcal{R}$ meets 8-1-(i,ii) are trivial. 8-22 obviously defines a linear operator L: $\mathcal{R}\to\cap_{p<\infty}L^p$ (compute explicitely $L\Phi$ in terms of Φ'_i, Φ''_i by 8-3 to check that $L\Phi\in\cap_{p<\infty}L^p$).

That L meets 8-3 and that Γ, defined by 8-2, meets 8-24, are also consequences of easy computations. 8-24 implies $\Gamma(\Phi,\Phi)\geq 0$, hence $(L,\mathcal{R})$ meets 8-1-(iv). Finally if $\Phi,\Psi\in\mathcal{R}$, property 8-1-(iii) for L' and L'' yields

$$\begin{aligned}E(\Phi L\Psi) &= \int P''(d\omega'')E'(\Phi'_{\omega''}\ L'\Psi'_{\omega''}) + \int P'(d\omega')E''(\Phi''_{\omega'}\ L''\Psi''_{\omega'})\\ &= \int P''(d\omega'')E'(\Psi'_{\omega''}\ L'\Phi'_{\omega''}) + \int P'(d\omega')E''(\Psi''_{\omega'}\ L''\Phi''_{\omega'}) = E(\Psi L\Phi).\end{aligned}$$

8-25 <u>PROPOSITION</u>: *a) Let* $\Phi':\Omega' \to \mathbb{R}$. *Then* $\Phi'\in H_2'$ *if and only if* $\Phi'\in H_2$, *in which case* $L'\Phi'$ *is a version of* $L\Phi'$.

b) If $\Phi'\in H_2'$, $\Phi''\in H_2''$, *then* $\Gamma(\Phi',\Phi'') = 0$.

As we shall see, the necessary part of (a), and (b), are almost evident; the non-trivial part is the sufficient condition in (a). We begin with an auxiliary lemma.

8-26 <u>LEMMA</u>: *Let* $\Phi\in\mathcal{R}$ *and* $\hat{\Phi}'(\omega')=E''(\Phi''_{\omega'})$. *Then* $\hat{\Phi}'\in\mathcal{R}'$ *and*

$$L'\hat{\Phi}'(\omega') = E''[(L\Phi)''_{\omega'}]. \tag{8-27}$$

<u>Proof</u>. We have $\Phi=F(\tilde{\Phi}',\tilde{\Phi}'')$, where $\tilde{\Phi}'=(\Phi_1',\ldots,\Phi_n')\in\mathcal{R}'^n$ and $\tilde{\Phi}''=(\Phi_1'',\ldots,\Phi_d'')\in\mathcal{R}''^d$ and $F\in C_p^2(R^n\times R^d)$. The function

$$G(x) = E''[F(x,\tilde{\Phi}'')] \tag{8-28}$$

clearly belongs to $C_p^2(R^n)$, and $\hat{\Phi}'=G(\tilde{\Phi}')$, hence $\hat{\Phi}'\in\mathcal{R}'$. We deduce from 8-22 and 8-5 that

$$\begin{aligned} E''[(L\Phi)''_{\omega'}] &= \int P''(d\omega'')L'\Phi'_{\omega''}(\omega') + E''(L''\Phi''_{\omega'}) \\ &= \int P''(d\omega'')L'\Phi'_{\omega''}(\omega'). \end{aligned}$$

Then, since we can differentiate 8-28 twice under the expectation, a simple computation shows that 8-27 holds (use 8-3 for L').

<u>Proof of 8-25</u>. a) First assume that $\Phi'\in H_2'$. There is a sequence $\{\Phi_n'\}\subset\mathcal{R}'$ such that $\|\Phi_n'-\Phi'\|_{H_2'} \to 0$. Clearly $\Phi_n'\in\mathcal{R}$, and $\Phi_n' \to \Phi'$ and $L'\Phi_n' \to L'\Phi'$ in $L^2(P)$ as well as in $L^2(P')$. Since $L'\Phi_n'=L\Phi_n'$ by 8-22 (recall that $L''1=0$), we obtain $\Phi'\in H_2$ and $L\Phi'=L'\Phi'$.

Conversely, assume that $\Phi'\in H_2$. There is a sequence

$\{\Phi_n\}\subset\mathcal{R}$ such that $\|\Phi_n-\Phi\|_{H_2'}\to 0$. Set $\hat{\Phi}_n'(\omega')=E''[(\Phi_n)''_{\omega'}]$. Then $\hat{\Phi}_n'\in\mathcal{R}'$ by the previous lemma, and

$$E'(|\hat{\Phi}_n'-\Phi'|^2) = \int P'(d\omega')\left|\int P''(d\omega'')[\Phi_n(\omega',\omega'')-\Phi'(\omega')]\right|^2$$

$$\leq \int P'(d\omega')\int P''(d\omega'')|\Phi_n(\omega',\omega'')-\Phi'(\omega')|^2$$

$$= E(|\Phi_n-\Phi'|^2) \to 0.$$

Since $\hat{\Phi}_n'$ satisfies 8-27, we also have

$$E'(|L'\hat{\Phi}_n'-L'\hat{\Phi}_m'|^2)$$

$$= \int P'(d\omega')\left|\int P''(d\omega'')[L\Phi_n(\omega',\omega'')-L\Phi_m(\omega',\omega'')]\right|^2$$

$$\leq \int P'(d\omega')\int P''(d\omega'')|L\Phi_n(\omega',\omega'')-L\Phi_m(\omega',\omega'')|^2$$

$$= E(|L\Phi_n-L\Phi_m|^2) \to 0$$

as $n,m\uparrow\infty$. Thus $\{L'\hat{\Phi}_n'\}$ is Cauchy in $L^2(P')$, thus $\Phi'\in H_2'$.

b) Let $\{\Phi_n'\}\subset\mathcal{R}'$ and $\{\Phi_n''\}\subset\mathcal{R}''$ with $\|\Phi_n'-\Phi'\|_{H_2'}\to 0$ and $\|\Phi_n''-\Phi''\|_{H_2''}\to 0$. From (a), we deduce that $\Phi_n'\to\Phi'$ and $\Phi_n''\to\Phi''$ in H_2, hence 8-13 yields $\Gamma(\Phi_n',\Phi_n'')\to\Gamma(\Phi',\Phi'')$ in $L^1(P)$. However, 8-24 implies that $\Gamma(\Phi_n',\Phi_n'')=0$, so $\Gamma(\Phi',\Phi'')=0$.

Section 9: MALLIAVIN OPERATOR ON WIENER-POISSON SPACE

§9-a. MALLIAVIN OPERATOR ON POISSON SPACE

In this subsection, $(\Omega,\underline{\underline{F}},P)$ denotes the canonical space introduced in §6-a, except that we have only the *Poisson measure* μ, and no Wiener process (recall that the intensity measure of μ is $\nu(dt,dz)=dt\times G(dz)$, G being Lebesgue measure on the open subset E of $\mathbb{R}^\beta$).

We denote by $C^2_{o,E}([0,T]\times E)$ the set of all functions $f:[0,T]\times E\to\mathbb{R}$ that are Borel, null outside a compact subset, of class C^2 on E (i.e., in the second variable) with $f, D_z f, D^2_{z^2} f$ uniformly bounded on $[0,T]\times E$. If $f\in C^2_{o,E}([0,T]\times E)$, we write $\mu(f)$ for the random variable $\int\mu(.;dt,dz)f(t,z)$.

We consider an *auxiliary function* $\rho:E\to[0,\infty)$ which is of class C^1_b (other conditions, similar to 6-9, will be put on ρ later). Set

$$\left.\begin{array}{l} \mathcal{R} = \text{the set of all functions of the form} \\ \Phi=F(\mu(f_1),\ldots,\mu(f_k)),\ \text{with}\ F\in C^2_p(R^k), \\ f_i\in C^2_{o,E}([0,T]\times E). \end{array}\right\} \quad (9\text{-}1)$$

If Φ is as above, we set

$$L\Phi = \frac{1}{2}\sum_{i=1}^{k}\frac{\partial}{\partial x_i}F(\mu(f_1),\ldots,\mu(f_k))\,\mu(\rho\Delta_z f_i + D_z\rho(D_z f_i)^T)$$
$$+\ \frac{1}{2}\sum_{i,j=1}^{k}\frac{\partial^2}{\partial x_i\partial x_j}F(\mu(f_1),\ldots,\mu(f_k))\,\mu(\rho D_z f_i(D_z f_j)^T) \qquad (9\text{-}2)$$

where Δ_z stands for the Laplacian on E.

9-3 PROPOSITION: *9-1 and 9-2 define a Malliavin operator* $(L,\mathcal{R})$. *Moreover if* $\Phi=F(\mu(f_1),..,\mu(f_k))$ *and* $\Psi=H(\mu(h_1),..,\mu(h_q))$ *belong to* $\mathcal{R}$, *then*

$$\Gamma(\Phi,\Psi) = \sum_{i=1}^{k}\sum_{j=1}^{q} \frac{\partial}{\partial x_i}F(\mu(f_1),..,\mu(f_k)) \frac{\partial}{\partial x_j}H(\mu(h_1),..,\mu(h_q))\ \mu(\rho D_z f_i (D_z h_j)^T). \tag{9-4}$$

Proof. a) It is well known that $\mu(f)\in\cap_{p<\infty}L^p$ whenever f is a bounded function on $[0,T]\times E$ with compact support. Hence $\mathcal{R}\subset\cap_{p<\infty}L^p$ and $L\Phi\in\cap_{p<\infty}L^p$ for $\Phi\in\mathcal{R}$. That $\mathcal{R}$ meets 8-1-(i,ii) is evident.

b) Next, we need to show that 9-2 defines a linear map on $\mathcal{R}$. In fact, the linearity will be evident, provided we prove the following: let $\Phi=F(\mu(f_1),..,\mu(f_k))$ and $\Psi=H(\mu(h_1),..,\mu(h_q))$ and define $L\Phi$ and $L\Psi$ by 9-2; then, if $\Phi\equiv\Psi$, we have $L\Phi\equiv L\Psi$.

Let $\mu\in\Omega$ be fixed. Let (t,z) be a point in the support of μ. Set $\mu'=\mu-\varepsilon_{(t,z)}$ and $\mu^\theta=\mu'+\varepsilon_{(t,z+\theta)}$ for $\theta\in V$, where V is a neighbourhood of 0 in $\mathbb{R}^\beta$ such that $z+V\in E$. Then $\mu^0=\mu$, and the function

$$\phi_{t,z}(\theta):=\Phi(\mu^\theta) = F(\mu'(f_1)+f_1(t,z+\theta),...,\mu'(f_k)+f_k(t,z+\theta))$$

is of class C^2 on V. Hence $\hat{\phi}_{t,z}(\theta):=\rho(z+\theta)D_\theta\phi_{t,z}(\theta)$ is a C^1 function from V into $\mathbb{R}^\beta$, and a simple computation shows that

$$\mathrm{div}_\theta\, \hat{\phi}_{t,z}(0) =$$

$$\sum_{i=1}^{k} \frac{\partial F}{\partial x_i}(\mu(f_1),\ldots)\{\rho(z)\Delta_z f_i(t,z)+D_z f_i(t,z)D_z\rho(z)^T\}$$
$$+ \sum_{i,j=1}^{k} \frac{\partial^2 F}{\partial x_i \partial x_j}(\mu(f_1),\ldots)\rho(z)D_z f_i(z,t)D_z f_j(z,t)^T.$$

Similarly, we associate $\psi_{t,z}$ and $\hat{\psi}_{t,z}$ with Ψ. Since $\Phi\equiv\Psi$ we deduce that $\mathrm{div}_\theta\hat{\phi}_{t,z}(0) = \mathrm{div}_\theta\hat{\psi}_{t,z}(0)$. Summing these equalities on all points (t,z) in the support of μ yields $L\Phi(\mu) = L\Psi(\mu)$, thus $L\Phi\equiv L\Psi$.

c) Two simple computations show that L meets 8-3, and that the associated operator Γ (as in 8-2) is given by 9-4. Since $\rho\geq 0$ and since the matrix $(c^{ij}= D_z f_i(D_z f_j)^T)_{i,j\leq k}$ is always symmetric nonnegative, we deduce from 9-4 that $\Gamma(\Phi,\Phi)\geq 0$: hence 8-1-(iv) holds.

d) It remains to prove 8-1-(iii). Let $\Phi = F(\mu(f_1),\ldots,\mu(f_k))$ and $\Psi=H(\mu(h_1),\ldots\mu(h_q))$ be in $\mathcal{R}$. Let K be a compact set of E such that $[0,T]\times K$ contains the supports of all f_i's and h_i's. Let $(S_i,Z_i)_{i\leq N}$ denote the points of $[0,T]\times K$ that belong to the support of μ, with $S_1<\ldots<S_N$ (if there is no such point, then $N=0$). One knows that N is a Poisson r.v. with parameter $TG(K)$, and that conditionally on the σ-field $\underline{\underline{G}}$ generated by N and $S_1,\ldots,S_N$, the r.v. $(Z_i)_{i\leq N}$ are independent with uniform distribution over K.

We will prove that $E(\Phi L\Psi+\frac{1}{2}\Gamma(\Phi,\Psi)\,|\,\underline{\underline{G}})=0$: this yields $E(\Phi L\Psi)=-\frac{1}{2}E(\Gamma(\Phi,\Psi))$, and we deduce 8-1-(iii) by symmetry between Φ and Ψ.

From now on, we fix N and the S_j's. For simplicity of notation, we set $f_i^j(z)=f_i(S_j,z)$ and $h_i^j(z)= h_i(S_j,z)$. From 9-2 and 9-4 we obtain that $\Phi L\Psi+ \frac{1}{2}\Gamma(\Phi,\Psi)= g(Z_1,\ldots,Z_N)$, where

$g(z_1,\ldots,z_N) =$

$$\frac{1}{2}F\left(\sum_1^N f_1^n(z_n),\ldots,\sum_1^N f_k^n(z_n)\right)\left\{\sum_{i,j=1}^{q}\frac{\partial^2 H}{\partial x_i \partial x_j}\left(\sum_1^N h_1^n(z_n),\ldots\right)\right.$$

$$\sum_{n=1}^{N}\rho(z_n)D_z h_i^n(z_n)D_z h_j^n(z_n)^T$$

$$+\sum_{i=1}^{q}\frac{\partial H}{\partial x_i}\left(\sum_1^N h_1^n(z_n),\ldots\right)\sum_{n=1}^{N}[\rho(z_n)\Delta_z h_i^n(z_n)$$

$$\left.+D_z\rho(z_n)D_z h_i^n(z_n)^T]\right\}$$

$$+\frac{1}{2}\sum_{j=1}^{k}\sum_{i=1}^{q}\frac{\partial F}{\partial x_j}\left(\sum_1^N f_1^n(z_n),\ldots\right)\frac{\partial H}{\partial x_i}\left(\sum_1^N h_1^n(z_n),\ldots\right)$$

$$\sum_{n=1}^{N}\rho(z_n)D_z f_j^n(z_n)D_z h_i^n(z_n)^T.$$

If G_K denotes the uniform distribution over K, it suffices to prove that

$$\int G_K(dz_1)\ldots G_K(dz_N)\; g(z_1,\ldots,z_N) = 0. \tag{9-5}$$

Let $n\leq N$ and $\hat{z}_n=\{z_j\}_{1\leq j\leq N, j\neq n}$. If $\ell\leq\beta$ we set

$$Q_{\hat{z}_n}^{n,\ell}(z_n) = F\left(\sum_{s=1}^{N} f_1^s(z_s),\ldots,\sum_{s=1}^{N} f_k^s(z_s)\right)$$

$$\times\sum_{i=1}^{q}\frac{\partial H}{\partial x_i}\left(\sum_1^N h_1^s(z_s),\ldots,\sum_1^N h_q^s(z_s)\right)\rho(z_n)\frac{\partial}{\partial x_\ell}h_i^n(z_n).$$

Then a simple computation shows that

$$g(z_1,\ldots,z_N) = \frac{1}{2}\sum_{n=1}^{N}\sum_{\ell=1}^{\beta}\frac{\partial}{\partial x_\ell}Q_{\hat{z}_n}^{n,\ell}(z_n).$$

Since $Q_{\hat{z}_n}^{n,\ell}=0$ on the boundary of K, it is then easy to deduce that

$$\int G_K(dz)\;\frac{\partial}{\partial x_\ell}Q_{\hat{z}_n}^{n,\ell}(z) = 0$$

for all n,ℓ,z_n: therefore 9-5 follows from 9-6.

§9-b. MALLIAVIN OPERATOR ON WIENER SPACE

Here we just recall, without proofs, how the proper Malliavin operator is defined on Wiener space. As a matter of fact, we give only one among many possible definitions proposed by various authors (see e.g. Stroock [24], Zakai [29]). It is very close to the definition proposed by Shigekawa [23] (see also [12]), and in fact even simpler.

Now, $(\Omega,\underline{\underline{F}},P)$ is the canonical m-dimensional Wiener space (the one introduced in §6-a, except that we only have the *Wiener process* W, and no Poisson measure). Set

$$\left.\begin{array}{l} \mathcal{R} = \text{the set of all functions of the form} \\ \Phi=F(W^{k_1}_{t_1},\dots,W^{k_n}_{t_n}),\ \text{where } F\in C^2_p(R^n), \\ 0\leq t_i\leq T,\ k_i\in\{1,2,\dots,m\}. \end{array}\right\} \quad (9\text{-}7)$$

If Φ is as above, and if $\delta_{ij}=0$ (resp. $=1$) if $i\neq j$ (resp. $i=j$), we set

$$\begin{aligned} L\Phi = &-\frac{1}{2}\sum_{i=1}^{n}\frac{\partial F}{\partial x_i}(W^{k_1}_{t_1},\dots)W^{k_i}_{t_i} \\ &+\frac{1}{2}\sum_{i,j=1}^{n}\frac{\partial^2 F}{\partial x_i\partial x_j}(W^{k_1}_{t_1},\dots)\delta_{k_i,k_j}(t_i\wedge t_j). \end{aligned} \quad (9\text{-}8)$$

9-9 PROPOSITION: *9-7 and 9-8 define a Malliavin operator* $(L,\mathcal{R})$. *Moreover, if* $\Phi=F(W^{k_1}_{t_1},\dots,W^{k_n}_{t_n})$ *and* $\Psi=H(W^{\ell_1}_{s_1},\dots,W^{\ell_q}_{s_q})$ *belong to* $\mathcal{R}$, *then*

$$\begin{aligned} \Gamma(\Phi,\Psi) = \sum_{i=1}^{n}\sum_{j=1}^{q}&\frac{\partial F}{\partial x_i}(W^{k_1}_{t_1},\dots) \\ &\frac{\partial H}{\partial x_j}(W^{\ell_1}_{s_1},\dots)\delta_{k_i,\ell_j}(t_i\wedge s_j). \end{aligned} \quad (9\text{-}10)$$

The proof is entirely elementary, except perhaps for 8-1-(iii) (which follows from a classical integration-by-parts argument for finite-dimensional Gaussian measures), and in any case is much simpler than for proving 9-3.

§9-c. MALLIAVIN OPERATOR ON WIENER-POISSON SPACE

From now on, $(\Omega,\underline{\underline{F}},P)$ is the canonical space of §6-a, with the Wiener process W and the Poisson measure μ. Then clearly,

$$(\Omega,\underline{\underline{F}},P) = (\Omega^P,\underline{\underline{F}}^P,P^P)\otimes(\Omega^W,\underline{\underline{F}}^W,P^W),$$

where $(\Omega^P,\underline{\underline{F}}^P,P^P)$ is the canonical Poisson space of §9-a and $(\Omega^W,\underline{\underline{F}}^W,P^W)$ is the canonical Wiener space of §9-b. Call $(L^P,\mathcal{R}^P)$ and $(L^W,\mathcal{R}^W)$ the above-constructed Malliavin operators on these spaces. Then we set

$$(L,\mathcal{R}) = \text{direct product of } (L^P,\mathcal{R}^P) \text{ and } (L^W,\mathcal{R}^W)\text{: see 8-21.} \tag{9-11}$$

With each these Malliavin operators we associate Γ,Γ^P,Γ^W and H_q, H_q^P, H_q^W as in Section 8. Then if $\Phi,\Psi\in H_2$ are respectively $\underline{\underline{F}}^P$- and $\underline{\underline{F}}^W$-measurable, we deduce from 8-25 that

$$\left.\begin{array}{l} L\Phi = L^P\Phi, \quad L\Psi = L^W\Psi, \quad \Gamma(\Phi,\Psi) = 0 \\ \Gamma(\Phi,\Phi) = \Gamma^P(\Phi,\Phi), \quad \Gamma(\Psi,\Psi) = \Gamma^W(\Psi,\Psi). \end{array}\right\} \tag{9-12}$$

We end this subsection with a property of adaptedness that is well known for $(L^W,\mathcal{R}^W)$. For $t\in[0,T]$, $\underline{\underline{F}}_t$ is defined in §6-a and we set

$$\underline{\underline{F}}^t = \sigma(W_s-W_t:s\geq t;\ \mu(A)\text{: A Borel subset of } (t,T]\times E). \tag{9-13}$$

9-14 PROPOSITION: *Let* $\Phi,\Psi\in H_2$ *be* $\underline{\underline{F}}_t$*-measurable. Then* $L\Phi$ *and* $\Gamma(\Phi,\Psi)$ *are* $\underline{\underline{F}}_t$*-measurable.*

b) Let $\Phi,\Psi\in H_2$ *with* Φ *being* $\underline{\underline{F}}_t$*-measurable, and* Ψ *being* $\underline{\underline{F}}^t$*-measurable. Then* $\Gamma(\Phi,\Psi)=0$.

Proof. Since $\underline{\underline{F}}_t$ is contained in the P-completion of $\underline{\underline{F}}_t^o=\sigma(W_s:s\leq t;\ \mu(A)$: A Borel subset of $[0,t]\times E)$, we can suppose Φ (and Ψ in case (a)) $\underline{\underline{F}}_t^o$-measurable. Call $(\Omega_I,\underline{\underline{F}}_I,P_I)$ the canonical Wiener-Poisson space when the time interval is an arbitrary interval I instead of $[0,T]$. We have the obvious identification:

$$(\Omega,\underline{\underline{F}},P) = (\Omega_{[0,t]},\underline{\underline{F}}_{[0,t]},P_{[0,t]}) \otimes(\Omega_{(t,T]},\underline{\underline{F}}_{(t,T]},P_{(t,T]})$$

and $\underline{\underline{F}}_{[0,t]}$ (resp. $\underline{\underline{F}}_{(t,T]}$) can be interpreted as the trace of $\underline{\underline{F}}_t^o$ (resp. $\underline{\underline{F}}^t$) over $\Omega_{[0,t]}$ (resp. $\Omega_{(t,T]}$). Moreover let $\mathcal{R}_t$ (resp. $\mathcal{R}^t$) be the set of all $\Phi\in\mathcal{R}$ that are measurable with respect to $\underline{\underline{F}}_t^o$ (resp. $\underline{\underline{F}}^t$), and let L_t (resp. L^t) be the restriction of L to $\mathcal{R}_t$ (resp. $\mathcal{R}^t$).

It easily follows from 9-11 and 9-1 and 9-3, and 9-7 and 9-8, that $(L_t,\mathcal{R}_t)$ and $(L^t,\mathcal{R}^t)$ are Malliavin operators on $(\Omega_{[0,t]},\underline{\underline{F}}_{[0,t]},P_{[0,t]})$ and on $(\Omega_{(t,T]},\underline{\underline{F}}_{(t,T]},P_{(t,T]})$, and that $(L,\mathcal{R})$ is the direct product of $(L_t,\mathcal{R}_t)$ and $(L^t,\mathcal{R}^t)$.

Now, a function Φ that is $\underline{\underline{F}}_t^o$- (resp. $\underline{\underline{F}}^t$-) measurable is just the extension to Ω of a measurable function on $(\Omega_{[0,t]},\underline{\underline{F}}_{[0,t]})$ (resp. $(\Omega_{(t,T]},\underline{\underline{F}}_{(t,T]})$). Hence, the claims readily follow from Proposition 8-25.

§9-d. MALLIAVIN OPERATORS AND STOCHASTIC INTEGRALS

Now we seek to recognize how L and Γ, as introduced in §9-c, operate on a (very particular) kind of stochastic integrals with respect to $\tilde{\mu}$ and W (we do not look here for a "general theory", which undoubtedly exists: see [16] for the Wiener case). More precisely, we consider two integrals of the form

$$\Psi = \int_t^T f(s,\Phi)dW_s \;,\quad \delta = \int_t^T\int_E F(s,z,\Phi)\tilde{\mu}(ds,dz) \qquad (9\text{-}15)$$

where:

9-16 (i) $\Phi=(\Phi^1,..,\Phi^d)$ is $\underline{\underline{F}}_t$-measurable and $\Phi^i \in H_\infty$ for all $1\le i\le d$;

(ii) $f:[0,T]\times\mathbb{R}^d \to \mathbb{R}^m$ is Borel, of class C^2 on $\mathbb{R}^d$, with $|D^r_{x^r} f(s,x)| \le \zeta(1+|x|^\theta)$ for $r=0,1,2$, for constants $\zeta,\theta\ge 0$;

(iii) $F:[0,T]\times E\times\mathbb{R}^d \to \mathbb{R}$ is Borel, of class C^2 on $E\times\mathbb{R}^d$, and there are two constants $\zeta,\theta\ge 0$ and a function $\eta\in\cap_{2\le p\le\infty} L^p(E,G)$ such that

$$|D^r_{x^r}F(s,z,x)| \le \zeta(1+|x|^\theta)\eta(z) \qquad \text{for } r=0,1,2$$

$$|D^r_{z^r}F(s,z,x)| \le \zeta(1+|x|^\theta) \qquad \text{for } r=1,2.$$

Note that in virtue of Lemma 5-1, the r.v. Ψ and δ above are well defined and belong to $\cap_{p<\infty} L^p$.

It is now the time to impose further restrictions to the C^1_b function ρ:

9-17 *Conditions on* ρ: If the complement E^c of E in $\mathbb{R}^\beta$ is not empty, and denoting by $d(z,E^c)$ the distance of $z\in E$ to E^c, the functions

$$z \leadsto \frac{\rho(z)}{d(z,E^c)^2 \wedge 1} \;, \qquad z \leadsto \frac{|D_z\rho(z)|}{d(z,E^c)\wedge 1}$$

belong to $\cap_{1\le p<\infty} L^p(E,G)$.

These conditions obviously imply 6-9-(ii) and also that $\rho, |D_z\rho| \in \cap_{1\le p\le\infty} L^p(E,G)$.

E

9-18 THEOREM: *Assume 9-16 and 9-17. Then the random variables* Ψ *and* δ *defined by 9-15 belong to* H_∞*, and we have 9-19 to 9-24 below:*

$$L\Psi = \int_t^T \{ \sum_{i=1}^d \frac{\partial f}{\partial x_i}(s,\Phi) L\Phi^i + \frac{1}{2} \sum_{i,j=1}^d \frac{\partial^2 f}{\partial x_i \partial x_j}(s,\Phi)\Gamma(\Phi^i,\Phi^j) - \frac{1}{2} f(x,\Phi)\} dW_s. \tag{9-19}$$

$$L\delta = \int_t^T \int_E \{ \sum_{i=1}^d \frac{\partial F}{\partial x_i}(s,z,\Phi) L\Phi^i + \frac{1}{2} \sum_{i,j=1}^d \frac{\partial^2 F}{\partial x_i \partial x_j}(s,z,\Phi)\Gamma(\Phi^i,\Phi^j)\} \tilde{\mu}(ds,dz) + \frac{1}{2} \int_t^T \int_E \{\rho(z)\Delta_z F(s,z,\Phi) + D_z\rho(z) D_z F(s,z,\Phi)^T\} d\mu. \tag{9-20}$$

$$\Gamma(\Psi,\Psi) = \int_t^T (ff^T)(s,\Phi)ds + \sum_{i,j=1}^d \Gamma(\Phi^i,\Phi^j)\{\int_t^T \frac{\partial f}{\partial x_i}(s,\Phi)dW_s\}\{\int_t^T \frac{\partial f}{\partial x_j}(s,\Phi)dW_s\}. \tag{9-21}$$

$$\Gamma(\delta,\delta) = \int_t^T \int_E \rho(z)(D_z F . D_z F^T)(s,z,\Phi)\mu(ds,dz) + \sum_{i,j=1}^d \Gamma(\Phi^i,\Phi^j)\{\int_t^T\int_E \frac{\partial F}{\partial x_i}(s,z,\Phi)d\tilde{\mu}\}\{\int_t^T\int_E \frac{\partial F}{\partial x_j}(s,z,\Phi)d\tilde{\mu}\}. \tag{9-22}$$

$$\Gamma(\Psi,\delta) = \sum_{i,j=1}^d \Gamma(\Phi^i,\Phi^j)\{\int_t^T \frac{\partial f}{\partial x_i}(s,\Phi)dW_s\} \{\int_t^T \int_E \frac{\partial F}{\partial x_j}(s,z,\Phi)d\tilde{\mu}\}. \tag{9-23}$$

If $\alpha \in H_\infty$, α *is* $\underline{\underline{F}}_t$*-measurable, then:*

$$\Gamma(\Psi,\alpha) = \sum_{i=1}^{d} \Gamma(\Phi^i,\alpha) \int_t^T \frac{\partial f}{\partial x_i}(s,\Phi)\,dW_s\,,$$

$$\Gamma(\delta,\alpha) = \sum_{i=1}^{d} \Gamma(\Phi^i,\alpha) \int_t^T\int_E \frac{\partial F}{\partial x_i}(s,z,\Phi)\,d\tilde{\mu}. \tag{9-24}$$

Of course, 9-21 and 9-22 allow to obtain $\Gamma(\Psi,\overline{\Psi})$ and $\Gamma(\delta,\overline{\delta})$ by polarization, if $\overline{\Psi}$ and $\overline{\delta}$ are defined by 9-15 with the functions $\overline{f}$ and $\overline{F}$ (meeting 9-16). Namely:

$$\Gamma(\Psi,\overline{\Psi}) = \int_t^T f(s,\Phi)\overline{f}(s,\Phi)^T ds \tag{9-25}$$

$$+ \sum_{i,j=1}^{d} \Gamma(\Phi^i,\Phi^j)\left\{\int_t^T \frac{\partial f}{\partial x_i}(s,\Phi)\,dW_s\right\}\left\{\int_t^T \frac{\partial \overline{f}}{\partial x_j}(s,\Phi)\,dW_s\right\}$$

$$\Gamma(\delta,\overline{\delta}) = \int_t^T\int_E \rho(z) D_z F(s,z,\Phi) D_z \overline{F}(s,z,\Phi)^T d\mu \tag{9-26}$$

$$+ \sum_{i,j=1}^{d} \Gamma(\Phi^i,\Phi^j)\left\{\int_t^T\int_E \frac{\partial F}{\partial x_i}(s,z,\Phi)\,d\tilde{\mu}\right\}\left\{\int_t^T\int_E \frac{\partial F}{\partial x_j}(s,z,\Phi)\,d\tilde{\mu}\right\}.$$

9-27 REMARK: Due to 9-16 and 5-1, all the integrals in the above formulae are well defined.

The proof of Theorem 9-18 will be broken into several steps. Firstly:

9-28 LEMMA: *Let* $f:[0,T] \to \mathbb{R}^m$ *be Borel and bounded. Then* $\Psi = \int_t^T f(s)\,dW_s$ *belongs to* H_∞, *and* $L\Psi = -\frac{1}{2}\int_t^T f(s)\,dW_s$ *and* $\Gamma(\Psi,\Psi) = \int_t^T f(s)f(s)^T ds$.

Proof. This result being well known in a slightly different context [16], we just sketch the proof. Firstly if f is a step function, then $\Psi\in\mathcal{R}$ and the formulae giving $L\Psi$ and $\Gamma(\Psi,\Psi)$ are nothing else than 9-8 and 9-10 (recall 9-12). If f is Borel and bounded, it is the limit in $L^2([0,T],dt)$ of a sequence (f_n) of step functions. Then $\Psi_n = \int_t^T f_n(s)dW_s \to \Psi$ in L^q for all $q<\infty$, and $L\Psi_n \to L\Psi=-\Psi/2$ in L^q, and $\Gamma(\Psi_n,\Psi_n) \to \Gamma(\Psi,\Psi):= \int_t^T f(s)f(s)^T ds$ in $L^{q/2}$: hence $\Psi\in H_{q/2}$ by 8-17.

Next, if B and K are compact subsets of E and $\mathbb{R}^d$:

$\mathcal{E}_{B\times K}$ = the set of all functions $F(s,z,x) = F_1(s,z)F_2(x)$, where $F_1\in C^2_{o,E}([0,T]\times E)$ has its support in $(t,T]\times E$, and $F_2\in C^2(R^d)$ has its support in K.

$\mathcal{E}'_{B\times K}$ = the linear space spanned by $\mathcal{E}_{B\times K}$.

$\mathcal{E}''_{B\times K}$ = the set of all functions F on $[0,T]\times E\times\mathbb{R}^d$ of class C^2 on $E\times\mathbb{R}^d$, with support in $(t,T]\times B\times K$, with $F, D_zF, D^2_{z^2}F, D_xF, D^2_{x^2}F$ uniformly bounded.

$\mathcal{F}_K$ = the set of all functions $f(s,x)=f_1(s)f_2(x)$ where f_1 is Borel bounded with support in $(t,T]$ and $f_2\in C^2(R^d)$ with support in K.

$\mathcal{F}'_K$ = the linear space spanned by $\mathcal{F}_K$.

$\mathcal{F}''_K$ = the set of all $\mathbb{R}^m$-valued functions f on $[0,T]\times\mathbb{R}^d$ of class C^2 on $\mathbb{R}^d$, with support in $(t,T]\times K$, with $f, D_xf, D^2_{x^2}f$ uniformly bounded.

9-29 <u>LEMMA</u>: *If* $f,\bar{f}\in\mathcal{F}'_K$ *and* $F,\bar{F}\in\mathcal{E}'_{B\times K}$ *we have* Ψ, $\bar{\Psi}$, δ, $\bar{\delta}\in H_\infty$ *and 9-19 to 9-26 hold.*

<u>Proof</u>. All equations 9-19 to 9-26 are linear, except 9-21 and 9-22, which indeed follow from 9-25 and 9-26 upon setting $\bar{f}=f$ and $\bar{F}=F$. So we can suppose that $f=f_1\otimes f_2$ and $\bar{f}=\bar{f}_1\otimes\bar{f}_2$ are in $\mathcal{F}_K$ and that $F=F_1\otimes F_2$ and $\bar{F}=\bar{F}_1\otimes\bar{F}_2$ are in $\mathcal{E}_{B\times K}$.

Set $\Psi'=\int_t^T f_1(s)dW_s$, $\bar{\Psi}'=\int_t^T \bar{f}_1(s)dW_s$, which are in H_∞ by 9-28. Set $\delta'=\mu(F_1)-\nu(F_1)$ and $\bar{\delta}'=\mu(\bar{F}_1)-\nu(\bar{F}_1)$, which are in H_∞ (and even in $\mathcal{R}$). By 8-18, $f_2(\Phi)$, $\overline{f}_2(\Phi)$, $F_2(\Phi)$ and $\overline{F}_2(\Phi)$ are in H_∞. Then $\Psi=\Psi' f_2(\Phi)$, $\overline{\Psi}=\overline{\Psi}'\overline{f}_2(\Phi)$, $\delta=\delta' F_2(\Phi)$ and $\overline{\delta}=\overline{\delta}'\overline{F}_2(\Phi)$ are all in H_∞.

We prove 9-20, for example. $\Gamma(\delta', F_2(\Phi))=0$ by 9-14, hence 8-3, 9-2 and 9-12 yield

$$\begin{aligned} L\delta &= \delta' L(F_2\circ\Phi) + F_2(\Phi)L\delta' \\ &= \tilde{\mu}(F_1)\Big\{\sum_{i=1}^{d}\frac{\partial}{\partial x_i}F_2(\Phi)L\Phi^i \\ &+\frac{1}{2}\sum_{i,j=1}^{d}\frac{\partial^2}{\partial x_i\partial x_j}F_2(\Phi)\Gamma(\Phi^i,\Phi^j)\Big\} + \frac{1}{2}F_2(\Phi)\mu(\rho\Delta_z F_1 + D_z\rho . D_z F_1^T), \end{aligned}$$

which is nothing else than 9-20: 9-19 is proved similarly (use 9-28). We also have

$$\begin{aligned} \Gamma(\delta,\overline{\delta}) &= \delta'\overline{\delta}'\Gamma(F_2(\Phi),\overline{F}_2(\Phi)) + F_2(\Phi)\overline{F}_2(\Phi)\Gamma(\delta',\overline{\delta}') \\ &= \sum_{i,j=1}^{d}\tilde{\mu}(F_1)\tilde{\mu}(\overline{F}_1)\frac{\partial}{\partial x_i}F_2(\Phi)\frac{\partial}{\partial x_j}\overline{F}_2(\Phi)\Gamma(\Phi^i,\Phi^j) \\ &\qquad + F_2(\Phi)\overline{F}_2(\Phi)\mu(\rho D_z F_1 . D_z\overline{F}_1^T) \end{aligned}$$

(use 9-4, 9-12, 9-14, 8-4), which is nothing else than 9-26; 9-25 is proved similarly. $\Gamma(\Psi',\delta')=0$ obtains from 9-12, hence

$$\begin{aligned} \Gamma(\Psi,\delta) &= \Psi'\delta'\Gamma(F_2(\Phi), f_2(\Phi)) \\ &= \sum_{i,j=1}^{d}\frac{\partial f_2}{\partial x_i}(\Phi)\frac{\partial F_2}{\partial x_j}(\Phi)\Gamma(\Phi^i,\Phi^j)\ \tilde{\mu}(F_1)\int_t^T f_1(s)dW_s, \end{aligned}$$

which is 9-23. Finally, 9-24 is proved similarly.

9-30 <u>LEMMA</u>: *Let* $F\in\mathcal{E}''_{B\times K}$: *let* $q\in\mathbb{N}^*$ *and* γ *be a positive finite measure on* $[0,T]\times E\times\mathbb{R}^d$; *let* B' *(resp.* K'*) be a compact neighbourhood of* B *(resp. of* K*). There is a sequence* $\{F_n\}\subset\mathcal{E}'_{B'\times K'}$ *such that* F_n, D_zF_n, $D^2_{z^2}F_n$, D_xF_n,

$D^2_{x^2}F_n$ *converge respectively to* F, D_zF, $D^2_{z^2}F$, D_xF, $D^2_{x^2}F$ *in* $L^q([0,T]\times E\times \mathbb{R}^d,\gamma)$.

<u>Proof</u>. We can obviously suppose that $E=\mathbb{R}^\beta$.

a) We first show that it suffices to prove the result when $F\in\mathcal{E}''_{B\times K}$ is C^∞ in (z,x). For this, let $B''\times K''$ be a compact neighbourhood of $B\times K$, contained in the interior of $B'\times K'$. Let $(\phi_n)\subset C^\infty_o(R^{\beta+d})$ be a regularizing sequence of functions, and set $G_n(s,.) = F(s,.)*\phi_n$ (the convolution product). Then G_n is C^∞ is (z,x), it belongs to $\mathcal{E}''_{B''\times K''}$ for all n large enough, and G_n and its first and second derivatives converge to the same for F, in $L^q(\gamma)$.

b) So we may suppose that F is C^∞ in (z,x). Set

$$G(s,.) = \frac{\partial^{2(\beta+d)}}{\partial z_1^2\ldots\partial z_\beta^2\partial x_1^2\ldots\partial x_d^2} F(s,.).$$

There is a sequence of Borel functions $G_n(s,z,x)= \sum_{1\leq i\leq k_n} g_{n,i}(s,z)\overline{g}_{n,i}(x)$, continuous in (z,x), such that $G_n\to G$ in $L^q(\gamma)$, with $g_{n,i}(s,z)=0$ if $s\leq t$. Now if h (resp. $\overline{h}$) is a C^∞ function on $\mathbb{R}^\beta$ (resp. $\mathbb{R}^d$) which equals 1 on B (resp. K) and 0 on B'^c (resp. K'^c) and if

$$F_n(s,z,x) = \sum_{i=1}^{k_n} h(z)\overline{h}(x)$$

$$\times\{\int_{\{u_i\leq z_i\}} du \int_{\{v_i\leq u_i\}} dv\ g_{n,i}(s,v)\}$$

$$\times\{\int_{\{y_i\leq x_i\}} dy \int_{\{y'_i\leq y_i\}} dy'\ \overline{g}_{n,i}(y')\},$$

(where $z=(z_i)$, $v=(v_i)$ are in $\mathbb{R}^\beta$, $y=(y_i)$, $y'=(y'_i)$ are in $\mathbb{R}^d$), the sequence (F_n) meets our claims.

There is of course a similar result about $''_K$, with a simpler proof.

9-31 LEMMA: *Let* $f \in F''_K$ *and* $F \in E''_{B\times K}$. *Then* δ *and* Ψ *belong to* H_∞, *and 9-19 to 9-24 hold.*

Proof. Let $B'\times K'$ be a compact neighbourhood of $B\times K$. Let γ_Φ be the law of Φ on $\mathbb{R}^d$, and $\nu' = 1_{(t,T]\times B'} \circ \nu$, and $\gamma'(ds) = 1_{(t,T]}(s)ds$, and set $\gamma = \nu' \otimes \gamma_\Phi$ and $\tilde{\gamma} = \gamma' \otimes \gamma_\Phi$ (finite measures on $[0,T]\times E\times \mathbb{R}^d$ and $[0,T]\times \mathbb{R}^d$). Let $q>4$ be fixed.

Let $\{F_n\} \subset E'_{B'\times K'}$ be a sequence approaching $F_\infty = F$ in $L^q([0,T]\times E\times\mathbb{R}^d, \gamma)$ in the sense of 9-30; let $\{f_n\} \subset F'_{K'}$ be a sequence approaching $f_\infty = f$ in $L^q([0,T]\times\mathbb{R}^d, \tilde{\gamma})$, again in the sense of 9-30.

For a function G_n on $[0,T]\times E\times\mathbb{R}^d$ (resp. g_n on $[0,T]\times\mathbb{R}^d$) we write $\mu(G_n) = \int_t^T \int_E G_n(s,z,\Phi)d\mu$ and similarly for $\nu(G_n)$ and $\tilde{\mu}(G_n)$ (resp. $w(g_n) = \int_t^T g_n(s,\Phi)dW_s$ and $\hat{w}(g_n) = \int_t^T g_n(s,\Phi)ds$), provided those random variables are well-defined. Then we deduce from 9-30 and from the very definition of γ and $\tilde{\gamma}$ that

9-32 $E(\nu(|G_n - G_\infty|^q)) \to 0$ (resp. $E(\hat{w}(|g_n - g_\infty|^q)) \to 0$) if G_n is one of the functions F_n, $\partial F_n/\partial x_i$, $\partial^2 F_n/\partial x_i \partial x_j$ or $\rho\Delta_z F_n + D_z\rho . D_z F_n^T$ (resp. if g_n is one of the functions f_n, $\partial f_n/\partial x_i$, or $\partial^2 f_n/\partial x_i \partial x_j$).

9-33 $E(\nu(|G_n - G_\infty|^{q/2})) \to 0$ (resp. $E(\hat{w}(|g_n - g_\infty|^{q/2})) \to 0$) if $G_n = \rho D_z F_n . D_z F_n^T$ (resp. $g_n = f_n f_n^T$).

Now, Hölder's inequality yields for $p \geq 1$:

$$E(|\mu(G_n) - \mu(G_\infty)|^p) \leq$$

$$\leq E[\mu((t,T]\times B')^{p-1/2}\ \mu(|G_n-G_\infty|^{2p})^{1/2}]$$

$$\leq E[\mu((t,T]\times B')^{2p-1}]^{1/2}\ E[\mu(|G_n-G_\infty|^{2p})]^{1/2}$$

$$= c_p\ E(\nu(|G_n-G_\infty|^{2p}))^{1/2}$$

(the last equality, where c_p is some constant, comes from the $\underline{\underline{P}}\otimes\underline{\underline{E}}$-measurability of $G_n(s,z,\Phi)$ in (ω,s,z)). Thus $\mu(G_n) \to \mu(G_\infty)$ in $L^{q/2}$ (resp. $L^{q/4}$) if G_n is like in 9-32 (resp. in 9-33), and by difference $\tilde{\mu}(G_n) \to \tilde{\mu}(G_\infty)$ in $L^{q/2}$ (resp. $L^{q/4}$) as well. We also deduce easily (from Lemma 5-1 for instance) that $w(g_n) \to w(g_\infty)$ in L^q (resp. $L^{q/2}$) if g_n is like in 9-32 (resp. in 9-33). Hence

$$\left.\begin{array}{l}
\Psi_n := w(f_n) \to \Psi := w(f),\\
\delta_n := \tilde{\mu}(F_n) \to \delta := \tilde{\mu}(F),\\
w(\frac{\partial f_n}{\partial x_i}) \to w(\frac{\partial f}{\partial x_i}),\ w(\frac{\partial^2 f_n}{\partial x_i\partial x_j}) \to w(\frac{\partial^2 f}{\partial x_i\partial x_j}),\\
\tilde{\mu}(\frac{\partial F_n}{\partial x_i}) \to \tilde{\mu}(\frac{\partial F}{\partial x_i}),\ \tilde{\mu}(\frac{\partial^2 F_n}{\partial x_i\partial x_j}) \to \tilde{\mu}(\frac{\partial^2 F}{\partial x_i\partial x_j}),\\
\mu(\rho\Delta_z F_n + D_z\rho . D_z F_n^T) \to \mu(\rho\Delta_z F + D_z\rho . D_z F^T),\\
\text{all these convergences taking place in } L^{q/2}
\end{array}\right\} \quad (9\text{-}34)$$

and

$$\left.\begin{array}{l}
\hat{w}(f_n f_n^T) \to \hat{w}(f f^T),\\
\mu(\rho D_z F_n . D_z F_n^T) \to \mu(\rho D_z F . D_z F^T)
\end{array}\right\} \text{ in } L^{q/4}. \quad (9\text{-}35)$$

We deduce from 9-29 that $\Psi_n, \delta_n \in H_\infty$ and we can write 9-19 to 9-24 for Ψ_n, δ_n, which gives for $n\in\mathbb{N}$:

$$L\Psi_n = \sum_{i=1}^{d} w(\frac{\partial f_n}{\partial x_i}) L\Phi^i - \frac{1}{2}\Psi_n + \frac{1}{2}\sum_{i,j=1}^{d} w(\frac{\partial^2 f_n}{\partial x_i \partial x_j})\Gamma(\Phi^i,\Phi^j),$$

$$\left.\begin{aligned}
L\delta_n &= \sum_{i=1}^{d} \tilde{\mu}\left(\frac{\partial F_n}{\partial x_i}\right) L\Phi^i + \frac{1}{2}\,\mu(\rho\Delta_z F_n + D_z\rho . D_z F_n^T) \\
&\quad + \frac{1}{2}\sum_{i,j=1}^{d} \tilde{\mu}\left(\frac{\partial^2 F_n}{\partial x_i \partial x_j}\right)\Gamma(\Phi^i,\Phi^j), \\
\Gamma(\Psi_n,\Psi_n) &= \hat{w}(f_n f_n^T) + \sum_{i,j=1}^{d} w\left(\frac{\partial f_n}{\partial x_i}\right) w\left(\frac{\partial f_n}{\partial x_j}\right)\Gamma(\Phi^i,\Phi^j) \\
\Gamma(\delta_n,\delta_n) &= \mu(\rho D_z F_n . D_z F_n^T) \\
&\quad + \sum_{i,j=1}^{d} \tilde{\mu}\left(\frac{\partial F_n}{\partial x_i}\right)\tilde{\mu}\left(\frac{\partial F_n}{\partial x_j}\right)\Gamma(\Phi^i,\Phi^j), \\
\Gamma(\Psi_n,\delta_n) &= \sum_{i,j=1}^{d} w\left(\frac{\partial f_n}{\partial x_i}\right)\tilde{\mu}\left(\frac{\partial F_n}{\partial x_j}\right)\Gamma(\Phi^i,\Phi^j), \\
\Gamma(\Psi_n,\alpha) &= \sum_{i=1}^{d} w\left(\frac{\partial f_n}{\partial x_i}\right)\Gamma(\Phi^i,\alpha), \\
\Gamma(\delta_n,\alpha) &= \sum_{i=1}^{d} \tilde{\mu}\left(\frac{\partial F_n}{\partial x_i}\right)\Gamma(\Phi^i,\alpha).
\end{aligned}\right\} \quad (9\text{-}36)$$

Moreover, if we define $L\Psi$, etc... by 9-19 to 9-24, those random variables are also given by 9-36, with f, F instead of f_n, F_n. Since $L\Phi^i$, $\Gamma(\Phi^i,\Phi^j)$, $\Gamma(\Phi^i,\alpha)\in\cap_{p<\infty}L^p$, we readily deduce from 9-34 and 9-35 that $L\Psi_n \to L\Psi$, $L\delta_n \to L\delta$ in $L^{q/2-\varepsilon}$ and $\Gamma(\Psi_n,\Psi_n) \to \Gamma(\Psi,\Psi)$, $\Gamma(\delta_n,\delta_n) \to \Gamma(\delta,\delta)$, $\Gamma(\Psi_n,\delta_n) \to \Gamma(\Psi,\delta)$, $\Gamma(\Psi_n,\alpha) \to \Gamma(\Psi,\alpha)$ and $\Gamma(\delta_n,\alpha) \to \Gamma(\delta,\alpha)$ in $L^{q/4-\varepsilon}$, all this for arbitrary $\varepsilon>0$. Then 8-17 yields that Ψ and δ belong to $H_{q/2-\varepsilon}$ and that 9-19 to 9-24 are valid. Since q is arbitrarily large, we obtain the result.

<u>Proof of Theorem 9-18</u>. For simplicity, we set $d(z)= d(z,E^c)\wedge 1$, where $d(z,E^c)$ is the distance between $z\in E$ and the complement E^c of E in $\mathbb{R}^\beta$ (and $d(z)\equiv 1$ if $E=\mathbb{R}^\beta$). For each $n\in\mathbb{N}^*$ we set

$$K_n = \{x\in\mathbb{R}^d : |x|\le n\}, \qquad C_n = \{z\in E: |z|\le n\},$$
$$D_n = \{z\in E: d(z)\ge\frac{1}{n}\}, \qquad B_n = C_n\cap D_n.$$

K_n and B_n are compact. Observing that in case $E^c\neq\emptyset$ the distance between D_n and $(D_{2n})^c$ is $1/2n$, it is easy to obtain functions $\phi_n:\mathbb{R}^d \to [0,1]$ and $\psi_n:E \to [0,1]$ which are C^2 and satisfy

$$\phi_n(x)=\begin{cases}1 & \text{if } x\in K_n\\ 0 & \text{if } x\notin K_{2n},\end{cases} \qquad \psi_n(z)=\begin{cases}1 & \text{if } z\in B_n\\ 0 & \text{if } z\notin B_{2n},\end{cases} \tag{9-37}$$

and

$$|D_x\phi_n| \le c, \quad |D^2_{x^2}\phi_n| \le c,$$
$$|D_z\psi_n(z)| \le c(1 + n1_{D_{2n}\setminus D_n}(z)) \le c(1+\frac{1}{d(z)}) \tag{9-38}$$
$$|D^2_{z^2}\psi_n(z)| \le c(1+n^2 1_{D_{2n}\setminus D_n}(z)) \le c(1+\frac{1}{d(z)^2})$$

for some constant c not depending on n. Then we set

$$f_n(s,x) = f(s,x)\phi_n(x)$$
$$F_n(s,z,x) = F(s,z,x)\psi_n(z)\phi_n(x). \tag{9-39}$$

By construction, $f_n\in\mathcal{F}_{K_{2n}}$ and $F_n\in\mathcal{E}''_{B_{2n}\times K_{2n}}$, so if we define Ψ_n and δ_n by 9-15, with f_n and F_n instead of f and F, Lemma 9-31 applies. In particular, $\Psi_n,\delta_n\in H_\infty$ and, with the notation of the proof of 9-31 (namely, $\mu(G_n)$, $\tilde{\mu}(G_n)$, $w(g_n)$, $\hat{w}(g_n)$), equations 9-36 are valid. If we can show that 9-34 and 9-35 hold for all $q<\infty$, the end of the proof of 9-31 will give the claimed results.

It thus remain to prove 9-34 and 9-35. Let θ and $\eta(z)$ be as in 9-16; since $d(z)\le 1$ and $\sup_z \eta(z)<\infty$, one easily deduces from 9-16, 9-38 and 9-39 that, for some constant ζ' (depending on ζ,c,η),

$$\left.\begin{aligned} |D^r_{x^r} f_n(s,x)| &\leq \zeta'(1+|x|^\theta), \quad r=0,1,2 \\ |D^r_{x^r} F_n(s,z,x)| &\leq \zeta'(1+|x|^\theta)\eta(z), \quad r=0,1,2 \\ |D^r_{z^r} F_n(s,z,x)| &\leq \zeta'(1+|x|^\theta), \quad r=1,2. \end{aligned}\right\} \quad (9\text{-}40)$$

Let also $H_s = \zeta'(1+|\Phi|^\theta)1_{(t,T]}(s)$, which is predictable and has:

$$\int_0^T E(|H_s|^p)ds < \infty \quad \text{for all } p<\infty. \qquad (9\text{-}41)$$

Let g_n be f_n or $\partial f_n/\partial x_i$ or $\partial^2 f_n/\partial x_i \partial x_j$, and accordingly for g. Then 9-37 and 9-39 yield $g_n \to g$ pointwise, while 9-40 yields $|g_n(s,\Phi)| \leq H_s$. Then one readily deduces from 9-41, 5-1-b and from Lebesgue dominated convergence theorem that $w(g_n) \to w(g)$ in L^q for all $q<\infty$. Similarly if $g_n = f_n f_n^T$ we obtain $\hat{w}(g_n) \to \hat{w}(g)$ in L^q for all $q<\infty$.

Let G_n be F_n or $\partial F_n/\partial x_i$ or $\partial^2 F_n/\partial x_i \partial x_j$, and accordingly for G. Then 9-37 and 9-39 yield $G_n \to G$ pointwise and 9-40 yields $|G_n(s,z,\Phi)| \leq H_s \eta(z)$. Then 9-41, 5-1-c and Lebesgue theorem again give that $\tilde{\mu}(G_n) \to \tilde{\mu}(G)$ in L^q for all $q<\infty$. Finally let $G_n = \rho \Delta_z F_n + D_z \rho . D_z F_n^T$ or $G_n = \rho D_z F_n . D_z F_n^T$, and accordingly for G. Here again, $G_n \to G$ pointwise. Moreover 9-40 yields $|G_n(s,z,\Phi)| \leq (H_s + H_s^2)\eta'(z)$, where $\eta'(z) = \rho(z)/d^2(z) + |D_z \rho(z)|/d(z)$ belongs to $\cap_{1 \leq p<\infty} L^p(E,G)$ in virtue of 9-17. This implies that $\eta' * \mu_T^* \in \cap_{p<\infty} L^p$, and in particular $\eta' * \mu_T < \infty$ a.s. Hence Lebesgue theorem yields $\mu(G_n) \to \mu(G)$ a.s. But

$$E(|\mu(G_n)|^p) \leq E[\sup_{s\leq T} |H_s + H_s^2|^p (\eta' * \mu_T)^p] < \infty,$$

so indeed $\mu(G_n) \to \mu(G)$ in L^q for all $q<\infty$.

At this point, we have proved that 9-34 and 9-35 hold for all $q<\infty$, so we are finished.

Section 10: MALLIAVIN OPERATOR AND STOCHASTIC DIFFERENTIAL EQUATIONS

§10-a. THE MAIN RESULT

In this whole Section 10, the setting is the same as in §6-a. The auxiliary function $\rho: E \to [0,\infty)$ meets:

10-1 *Condition on* ρ: (i) ρ is of class C_b^∞, with $|D^r_{z^r}\rho| \in L^1(E,G)$ for all $r \in \mathbb{N}$ (the same as 6-9-(i)).

(ii) Condition 9-17 is met (recall that this is stronger than 6-9-(ii)).

With this function, the Malliavin operator $(L,\mathcal{R})$ is as defined in §9-c.

We consider Equation 6-2, with a fixed initial condition x_0. We assume something slightly stronger than (A'-r):

10-2 ASSUMPTION $(\tilde{A}'\text{-}r)$: *The same as (A'-r) (see 6-3), except that* $\eta \in \cap_{2\le p\le\infty} L^p(E,G)$.

Our aim in this subsection is to prove the:

10-3 THEOREM: *Assume* $(\tilde{A}'\text{-}3)$ *and 10-1. Then for each* $t \le T$ *the components* X^i_t *of the solution* X *to 6-2 belong to* H_∞. *Moreover if* $U^{ij}_t = \Gamma(X^i_t, X^j_t)$ *and* $V^i_t = LX^i_t$, *there are versions of the processes* $U=(U^{ij})_{i,j\le d}$ *and* $V=(V^i)_{i\le d}$ *that are solutions to the following linear equations:*

$$U = bb^T(X_-)*t + \rho D_z c(X_-) D_z c(X_-)^T * \mu$$
$$+ [U_- D_x a(X_-)^T + D_x a(X_-) U_-] * t$$
$$+ \sum_{i=1}^{m} [U_- D_x b^{\cdot i}(X_-)^T + D_x b^{\cdot i}(X_-) U_-] * W^i \tag{10-4}$$
$$+ [U_- D_x c(X_-)^T + D_x c(X_-) U_-] * \tilde{\mu}$$
$$+ \sum_{i=1}^{m} D_x b^{\cdot i}(X_-) U_- D_x b^{\cdot i}(X_-)^T * t$$
$$+ D_x c(X_-) U_- D_x c(X_-)^T * \mu,$$

$$V = \frac{1}{2} \sum_{i,j=1}^{d} \{ \frac{\partial^2 a}{\partial x_i \partial x_j}(X_-) U_-^{ij} * t$$
$$+ \frac{\partial^2 b}{\partial x_i \partial x_j}(X_-) U_-^{ij} * W + \frac{\partial^2 c}{\partial x_i \partial x_j}(X_-) U_-^{ij} * \tilde{\mu} \} \tag{10-5}$$
$$-\frac{1}{2} b(X_-) * W + \frac{1}{2}[\rho \Delta_z c(X_-) + D_z \rho D_z c(X_-)^T] * \mu$$
$$+ D_x a(X_-) V_- * t + D_x b(X_-) V_- * W + D_x c(X_-) V_- * \tilde{\mu}.$$

To prove this theorem, we begin by introducing the Peano's approximation to 6-2, namely

$$X^n = x_0 + a(X^n_{\phi^n}) * t + b(X^n_{\phi^n}) * W + c(X^n_{\phi^n}) * \tilde{\mu} \tag{10-6}$$

where ϕ^n is defined in 5-28. Then 6-2 and 10-6 are of type 5-3 and 5-27 respectively, with $A^n = A = a$, $B^n = B = b$, $C^n = C = c$. So assumption 5-29 is readily verified, with $Z_t'^n = 0$, and from Theorem 5-31 we deduce that 10-6 has a unique solution X^n, and that

$$\| (X^n - X)^*_T \|_{L^p} \to 0 \quad \text{for all } p < \infty. \tag{10-7}$$

Next, for each $n \in \mathbb{N}^*$ we define the following càdlàg process:

$$K_t^n = D_x a(X^n_{\phi_t^n})(t-\phi_t^n) + D_x b(X^n_{\phi_t^n})(W_t - W_{\phi_t^n})$$
$$+ \int_{\phi_t^n}^t \int_E D_x c(X^n_{\phi_t^n}, z)\tilde{\mu}(ds,dz). \tag{10-8}$$

10-9 LEMMA: *For every* $p\in[1,\infty)$ *there is a constant* δ_p, *not depending on* n, *such that for all* $t\leq T$:

$$E(|K^n_{t-}|^p) \leq \delta_p 2^{-n} \tag{10-10}$$

$$E(|K^{n,ij}_{t-}|^p|\underline{\underline{F}}_{\phi_t^n}) \leq \delta_p 2^{-n} \quad \text{for } M_{s-1}<i,j\leq M_s \tag{10-11}$$

(in 10-11, $M_s = d_1+..+d_s$, when $\mathbb{R}^d$ is graded as $\mathbb{R}^d = \mathbb{R}^{d_1}\times\ldots\times\mathbb{R}^{d_q}$ for the grading appearing in ($\tilde{A}$'-3)).

<u>Proof</u>. Let $\alpha^{ij}(x,z)$ be any of the functions $\partial a^i/\partial x_j(x)$, or $\partial b^{ik}/\partial x_j(x)$, or $\frac{1}{\eta(z)}\partial c^i/\partial x_j(x,z)$. By (i) of ($\tilde{A}$'-3) we have $|\alpha^{ij}(X^n_{\phi_t^n},z)|\leq\zeta(1+|X^n_{\phi_t^n}|^\theta)$. Then 10-8 and Lemma 5-1 yield a constant δ_p^1 independent on t and n, with

$$E(|K^{n,ij}_{t-}|^p) \leq \delta_p^1\zeta^p E[(1+|X^n_{\phi_t^n}|^\theta)^p](t-\phi_t^n).$$

We deduce $\sup_{n,t} E[(1+|X^n_{\phi_t^n}|^\theta)^p] <\infty$ from 10-7, while $t-\phi_t^n\leq T2^{-n}$. Then 10-10 follows.

If moreover i,j are like in 10-11, by (iv) of ($\tilde{A}$'-3) we have $|\alpha^{ij}(X^n_{\phi_t^n},z)|\leq\zeta$. Then if $B\in\underline{\underline{F}}_{\phi_t^n}$ we obtain from Lemma 5-1:

$$E(1_B|K^{n,ij}_{t-}|^p) \leq \delta_p^1\zeta^p\int_{\phi_t^n}^t E[(1_B)^p]ds = \delta_p^1\zeta^p P(B)(t-\phi_t^n).$$

This being true for all $B\in\underline{\underline{F}}_{\phi_t^n}$, we deduce 10-11 with $\delta_p = \delta_p^1\zeta^p T$.

10-12 <u>LEMMA</u>: *Let* $n\in\mathbb{N}^*$. *We have* $X_t^{n,i}\in H_\infty$ *for all* $i\leq d$, *and* $U_t^{n,ij} = \Gamma(X_t^{n,i}, X_t^{n,j})$ *and* $V_t^{n,i} = LX_t^{n,i}$ *are solutions of*

$$U^n = bb^T(X^n_{\phi^n}) * t + \rho D_z c(X^n_{\phi^n}) D_z c(X^n_{\phi^n})^T * \mu \qquad (10\text{-}13)$$

$$+ [U^n_{\phi^n} D_x a(X^n_{\phi^n})^T + D_x a(X^n_{\phi^n}) U^n_{\phi^n}] * t$$

$$+ \sum_{i=1}^m [U^n_{\phi^n} D_x b^{\cdot i}(X^n_{\phi^n})^T + D_x b^{\cdot i}(X^n_{\phi^n}) U^n_{\phi^n}] * W^i$$

$$+ [U^n_{\phi^n} D_x c(X^n_{\phi^n})^T + D_x c(X^n_{\phi^n}) U^n_{\phi^n}] * \tilde{\mu}$$

$$+ \sum_{i=1}^m D_x b^{\cdot i}(X^n_{\phi^n}) U^n_{\phi^n} D_x b^{\cdot i}(X^n_{\phi^n})^T * t$$

$$+ D_x c(X^n_{\phi^n}) U^n_{\phi^n} D_x c(X^n_{\phi^n})^T * \mu$$

$$+ [\underline{K}^n U^n_{\phi^n} D_x a(X^n_{\phi^n})^T + D_x a(X^n_{\phi^n}) U^n_{\phi^n} \underline{K}^{n,T}] * t$$

$$+ [\underline{K}^n U^n_{\phi^n} D_x b(X^n_{\phi^n})^T + D_x b(X^n_{\phi^n}) U^n_{\phi^n} \underline{K}^{n,T}] * W$$

$$+ [\underline{K}^n U^n_{\phi^n} D_x c(X^n_{\phi^n})^T + D_x c(X^n_{\phi^n}) U^n_{\phi^n} \underline{K}^{n,T}] * \tilde{\mu},$$

$$V^n = \frac{1}{2} \sum_{i,j=1}^d \{ \frac{\partial^2 a}{\partial x_i \partial x_j}(X^n_{\phi^n}) U^{n,ij}_{\phi^n} * t \qquad (10\text{-}14)$$

$$+ \frac{\partial^2 b}{\partial x_i \partial x_j}(X^n_{\phi^n}) U^{n,ij}_{\phi^n} * W + \frac{\partial^2 c}{\partial x_i \partial x_j}(X^n_{\phi^n}) U^{n,ij}_{\phi^n} * \tilde{\mu} \}$$

$$- \frac{1}{2} b(X^n_{\phi^n}) * W + \frac{1}{2}[\rho \Delta_z c(X^n_{\phi^n}) + D_z \rho D_z c(X^n_{\phi^n})^T] * \mu$$

$$+ D_x a(X^n_{\phi^n}) V^n_{\phi^n} * t + D_x b(X^n_{\phi^n}) V^n_{\phi^n} * W + D_x c(X^n_{\phi^n}) V^n_{\phi^n} * \tilde{\mu}.$$

<u>Proof</u>. Since $X_0^n = x_0$ we have $X_0^{n,i} \in H_\infty$ and $U_0^n = 0$, $V_0^n = 0$. Suppose that $X^{n,i}_{kT2^{-n}} \in H_\infty$ for some $k \in \mathbb{N}$ and all $i \leq d$. Let $t \in (kT2^{-n}, (k+1)T2^{-n}]$, hence $\phi_t^n = kT2^{-n}$, and 10-6 reads

$$X_t^{n,i} = X^{n,i}_{\phi_t^n} + a^i(X^n_{\phi_t^n})(t - \phi_t^n) + \int_{\phi_t^n}^t b^{i\cdot}(X^n_{\phi_t^n}) dW_s$$

$$+ \int_{\phi_t^n}^t \int_E c^i(X^n_{\phi_t^n}, z) \tilde{\mu}(ds, dz),$$

and call Ψ^i and δ^i the first and the second integrals above. By 8-18 we have $a^i(X^n_{\phi^n_t}) \in H_\infty$. Ψ^i and δ^i are given by 9-15, with $f(s,x)=b^{i\cdot}(x)1_{(\phi^n_t,t]}(s)$ and $F(s,z,x)=c^i(x,z)1_{(\phi^n_t,t]}(s)$. These functions meet 9-16 with ϕ^n_t instead of t by $(\tilde{A}'$-3). Then Theorem 9-18 yields that $\Psi^i, \delta^i \in H_\infty$: thus $X^{n,i}_t \in H_\infty$ for all $i \leq d$. This is true in particular for $t=(k+1)T2^{-n}$: then an induction on k shows that $X^{n,i}_t \in H_\infty$ for all $t \leq T$.

Next, using 8-3 and 9-19 and 9-20, we obtain

$$\begin{aligned} V^{n,i}_t = {} & V^{n,i}_{\phi^n_t} + \sum_{j=1}^d (t-\phi^n_t)\frac{\partial a^i}{\partial x_j}(X^n_{\phi^n_t})V^{n,j}_{\phi^n_t} \\ & + \frac{1}{2}\sum_{j,k=1}^d (t-\phi^n_t)\frac{\partial^2 a^i}{\partial x_j \partial x_k}(X^n_{\phi^n_t})U^{n,jk}_{\phi^n_t} \\ & + \int_{\phi^n_t}^t \{\sum_{j=1}^d \frac{\partial b^{i\cdot}}{\partial x_j}(X^n_{\phi^n_t})V^{n,j}_{\phi^n_t} + \frac{1}{2}\sum_{j,k=1}^d \frac{\partial^2 b^{i\cdot}}{\partial x_j \partial x_k}(X^n_{\phi^n_t})U^{n,jk}_{\phi^n_t} \\ & - \frac{1}{2}b^{i\cdot}(X^n_{\phi^n_t})\}dW_s + \int_{\phi^n_t}^t \int_E \{\sum_{j=1}^d \frac{\partial c^i}{\partial x_j}(X^n_{\phi^n_t},z)V^{n,j}_{\phi^n_t} \\ & + \frac{1}{2}\sum_{j,k=1}^d \frac{\partial^2 c^i}{\partial x_j \partial x_k}(X^n_{\phi^n_t},z)U^{n,jk}_{\phi^n_t}\}\tilde{\mu}(ds,dz) \\ & + \frac{1}{2}\int_{\phi^n_t}^t \int_E \{\rho(z)\Delta_z c^i(X^n_{\phi^n_t},z) + D_z\rho(z)D_z c^i(X^n_{\phi^n_t},z)^T\}\mu(ds,dz). \end{aligned}$$

Recalling that $V^n_0=0$, this easily gives 10-14.

Next, using 8-4 and 9-23, 9-24, 9-25 and 9-26, we obtain

$$\begin{aligned} U^{n,ij}_t = {} & U^{n,ij}_{\phi^n_t} + (t-\phi^n_t)\sum_{k=1}^d \{U^{n,ik}_{\phi^n_t}\frac{\partial a^j}{\partial x_k}(X^n_{\phi^n_t}) \\ & + U^{n,kj}_{\phi^n_t}\frac{\partial a^i}{\partial x_k}(X^n_{\phi^n_t})\} + (t-\phi^n_t)^2 \sum_{k,\ell=1}^d U^{n,k\ell}_{\phi^n_t}\frac{\partial a^i}{\partial x_k}\frac{\partial a^j}{\partial x_\ell}(X^n_{\phi^n_t}) \end{aligned}$$

$$+\int_{\phi_t^n}^{t} \sum_k \{U^{n,ik}_{\phi_t^n} \frac{\partial b^{j\cdot}}{\partial x_k}(X^n_{\phi_t^n}) + U^{n,kj}_{\phi_t^n} \frac{\partial b^{i\cdot}}{\partial x_k}(X^n_{\phi_t^n})\} dW_s$$

$$+ (t-\phi_t^n)\int_{\phi_t^n}^{t} \sum_{k,\ell} U^{n,k\ell}_{\phi_t^n} \{\frac{\partial a^i}{\partial x_k} \frac{\partial b^{j\cdot}}{\partial x_\ell}(X^n_{\phi_t^n}) + \frac{\partial a^j}{\partial x_k} \frac{\partial b^{i\cdot}}{\partial x_\ell}(X^n_{\phi_t^n})\} dW_s$$

$$+ \int_{\phi_t^n}^{t} b^{i\cdot}(b^{j\cdot})^T(X^n_{\phi_t^n}) ds$$

$$+ \sum_{k,\ell} \{\int_{\phi_t^n}^{t} \frac{\partial b^{i\cdot}}{\partial x_k}(X^n_{\phi_t^n}) dW_s\}\{\int_{\phi_t^n}^{t} \frac{\partial b^{j\cdot}}{\partial x_\ell}(X^n_{\phi_t^n}) dW_s\} U^{n,k\ell}_{\phi_t^n}$$

$$+ \sum_k \int_{\phi_t^n}^{t} \int_E \{U^{n,ik}_{\phi_t^n} \frac{\partial c^j}{\partial x_k}(X^n_{\phi_t^n}, z) + U^{n,kj}_{\phi_t^n} \frac{\partial c^i}{\partial x_k}(X^n_{\phi_t^n}, z)\} d\tilde{\mu}$$

$$+(t-\phi_t^n) \sum_{k,\ell} \int_{\phi_t^n}^{t}\int_E U^{n,k\ell}_{\phi_t^n} \{\frac{\partial a^i}{\partial x_k} \frac{\partial c^j}{\partial x_\ell}(X^n_{\phi_t^n}, z) + \frac{\partial a^j}{\partial x_k} \frac{\partial c^i}{\partial x}(X^n_{\phi_t^n}, z)\} d\tilde{\mu}$$

$$+ \sum_{k,\ell} \{\int_{\phi_t^n}^{t} \frac{\partial b^{i\cdot}}{\partial x_k}(X^n_{\phi_t^n}) dW_s \int_{\phi_t^n}^{t} \int_E \frac{\partial c^j}{\partial x_\ell}(X^n_{\phi_t^n}, z) d\tilde{\mu} + \int_{\phi_t^n}^{t} \frac{\partial b^{j\cdot}}{\partial x_k}(X^n_{\phi_t^n}) dW_s \int_{\phi_t^n}^{t} \int_E \frac{\partial c^i}{\partial x_\ell}(X^n_{\phi_t^n}, z) d\tilde{\mu}\} U^{n,k\ell}_{\phi_t^n}\}$$

$$+ \int_{\phi_t^n}^{t} \int_E \rho(z) D_z c^i D_z c^{j,T}(X^n_{\phi_t^n}, z) d\mu$$

$$+ \sum_{k,\ell=1}^{d} U^{n,k\ell}_{\phi_t^n} \{\int_{\phi_t^n}^{t} \int_E \frac{\partial c^i}{\partial x_k}(X^n_{\phi_t^n}, z) d\tilde{\mu}\}\{\int_{\phi_t^n}^{t} \int_E \frac{\partial c^j}{\partial x_\ell}(X^n_{\phi_t^n}, z) d\tilde{\mu}\}$$

So we have:

$$U^{n,ij}_t = U^{n,ij}_{\phi_t^n} + \int_{\phi_t^n}^{t} b^{i\cdot}(b^{j\cdot})^T(X^n_{\phi_t^n}) ds \tag{10-15}$$

$$+ \int_{\phi_t^n}^{t} \int_E \rho(z) D_z c^i D_z c^{j,T}(X^n_{\phi_t^n}, z) d\mu$$

$$+ \int_{\phi_t^n}^t \sum_k \{U_{\phi_t^n}^{n,ik} \frac{\partial a^j}{\partial x_k}(X_{\phi_t^n}^n) + U_{\phi_t^n}^{n,kj} \frac{\partial a^i}{\partial x_k}(X_{\phi_t^n}^n)\} ds$$

$$+ \int_{\phi_t^n}^t \sum_k \{U_{\phi_t^n}^{n,jk} \frac{\partial b^{i\cdot}}{\partial x_k}(X_{\phi_t^n}^n) + U_{\phi_t^n}^{n,ki} \frac{\partial b^{j\cdot}}{\partial x_k}(X_{\phi_t^n}^n)\} dW_s$$

$$+ \int_{\phi_t^n}^t \int_E \sum_k \{U_{\phi_t^n}^{n,ik} \frac{\partial c^j}{\partial x_k}(X_{\phi_t^n}^n, z) + U_{\phi_t^n}^{n,kj} \frac{\partial c^i}{\partial x_k}(X_{\phi_t^n}^n, z)\} d\tilde{\mu}$$

$$+ \sum_{k,\ell=1}^d U_{\phi_t^n}^{n,k\ell} K_t^{n,ik} K_t^{n,j\ell}$$

(recall that U^n is a symmetric matrix, and K^n is given by 10-8). Then Ito's formula yields

$$K_t^{n,ik} K_t^{n,j\ell} =$$

$$= \int_{\phi_t^n}^t \{K_{s-}^{n,ik} \frac{\partial a^j}{\partial x_\ell}(X_{\phi_t^n}^n) + K_{s-}^{n,j\ell} \frac{\partial a^i}{\partial x_k}(X_{\phi_t^n}^n)\} ds$$

$$+ \int_{\phi_t^n}^t \{K_{s-}^{n,ik} \frac{\partial b^{j\cdot}}{\partial x_\ell}(X_{\phi_t^n}^n) + K_{s-}^{n,j\ell} \frac{\partial b^{i\cdot}}{\partial x_k}(X_{\phi_t^n}^n)\} dW_s$$

$$+ \int_{\phi_t^n}^t \int_E \{K_{s-}^{n,ik} \frac{\partial c^j}{\partial x_\ell}(X_{\phi_t^n}^n, z) + K_{s-}^{n,j\ell} \frac{\partial c^i}{\partial x_k}(X_{\phi_t^n}^n, z)\} d\tilde{\mu}$$

$$+ \int_{\phi_t^n}^t \sum_r \frac{\partial b^{ir}}{\partial x_k} \frac{\partial b^{jr}}{\partial x_\ell}(X_{\phi_t^n}^n) ds + \int_{\phi_t^n}^t \int_E \frac{\partial c^i}{\partial x_k} \frac{\partial c^j}{\partial x_\ell}(X_{\phi_t^n}^n, z) d\mu.$$

Plugging this into 10-15 yields 10-13.

Before proving Theorem 10-3, we still need one more auxiliary result. We denote here by ∇X the process ∇X^{x_0} introduced in 6-30, and solution to

$$\nabla X = I + D_x(X_-)\nabla X_- * t + D_x b(X_-)\nabla X_- * W + D_x c(X_-)\nabla X_- * \tilde{\mu} \tag{10-16}$$

10-17 LEMMA: *Assume* $(\tilde{A}'\text{-}r)$ *for some* $r \geq 3$, *and call* U *and* V *the solutions to the linear equations 10-4 and 10-5. Then* $(X, \nabla X, U)$ *and* (X, U) *are solutions to equations of type 6-2 satisfying* $(\tilde{A}'\text{-}(r-1))$, *and* $(X, \nabla X, U, V)$ *and* (X, U, V) *are solutions to equations of type 6-2 satisfying* $(\tilde{A}'\text{-}(r-2))$.

Proof. Let $\overline{X} = (X, \nabla X, U, V)$, which takes its values in $F = \mathbb{R}^{\overline{d}}$, where $\overline{d} = 2d + d^2$; the points of F are denoted (x, y, u, v). Putting together 6-2, 10-16, 10-4 and 10-5, one sees that $\overline{X}$ satisfies an equation 6-2, with initial condition $(x_0, I, 0, 0)$ and coefficients $(\overline{a}, \overline{b}, \overline{c})$ given by:

10-18 If $i \leq d$, $\overline{a}^i(x,y,u,v) = a^i(x)$,
$\overline{b}^{is}(x,y,u,v) = b^{is}(x)$, $\overline{c}^i((x,y,u,v),z) = c^i(x,z)$.

10-19 If $d < i \leq d + d^2$ and i corresponds to the $(j,k)^{th}$ component of $\mathbb{R}^d \otimes \mathbb{R}^d$:

$$\overline{a}^i(x,y,u,v) = \sum_{\ell=1}^{d} \frac{\partial a^j}{\partial x_\ell}(x) y^{\ell k},$$

$$\overline{b}^{is}(x,y,u,v) = \sum_{\ell=1}^{d} \frac{\partial b^{js}}{\partial x_\ell}(x) y^{\ell k}$$

$$\overline{c}^i((x,y,u,v),z) = \sum_{\ell=1}^{d} \frac{\partial c^j}{\partial x_\ell}(x,z) y^{\ell k}.$$

10-20 If $d + d^2 < i \leq d + 2d^2$ and i corresponds to the $(j,k)^{th}$ component of $\mathbb{R}^d \otimes \mathbb{R}^d$:

$$\overline{a}^i(x,y,u,v) = \sum_{s=1}^{m} b^{js} b^{ks}(x)$$

$$+ \int_E \rho(z) \sum_{\ell=1}^{d} \frac{\partial c^j}{\partial x_\ell} \frac{\partial c^k}{\partial x_\ell}(x,z)\,dz$$

$$+ \sum_{\ell=1}^{d} \{u^{j\ell} \frac{\partial a^k}{\partial x_\ell}(x) + u^{\ell k} \frac{\partial a^j}{\partial x_\ell}(x)\} +$$

$$+ \sum_{\ell,r=1}^{d} \sum_{s=1}^{m} \frac{\partial b^{js}}{\partial x_\ell}(x) u^{\ell r} \frac{\partial b^{ks}}{\partial x_r}(x)$$

$$+ \int_E \sum_{\ell,r=1}^{d} \frac{\partial c^j}{\partial x_\ell}(x,z) u^{\ell r} \frac{\partial c^k}{\partial x_r}(x,z)\,dz \quad ,$$

$$\bar{b}^{is}(x,y,u,v) = \sum_{\ell=1}^{d} \{u^{j\ell} \frac{\partial b^{ks}}{\partial x_\ell}(x) + u^{\ell k} \frac{\partial b^{js}}{\partial x_\ell}(x)\}$$

$$\bar{c}^{i}((x,y,u,v),z) = \rho(z) \sum_{\ell=1}^{d} \frac{\partial c^j}{\partial x_\ell} \frac{\partial c^k}{\partial x_\ell}(x,z)$$

$$+ \sum_{\ell=1}^{d} \{u^{j\ell} \frac{\partial c^k}{\partial x_\ell}(x,z) + u^{\ell k} \frac{\partial c^j}{\partial x_\ell}(x,z)\}$$

$$+ \sum_{\ell,r=1}^{d} \frac{\partial c^j}{\partial x_\ell}(x,z) u^{\ell r} \frac{\partial c^k}{\partial x_r}(x,z) .$$

10-21 If $d+2d^2 < i \leq \bar{d}$ and $j=i-d-2d^2$,

$$\bar{a}^i(x,y,u,v) = \frac{1}{2} \sum_{\ell,r=1}^{d} \frac{\partial^2 a^j}{\partial x_\ell \partial x_r}(x) u^{\ell r} + \sum_{\ell=1}^{d} \frac{\partial a^j}{\partial x_\ell}(x) v^\ell$$

$$+ \frac{1}{2} \int_E \{\rho(z) \Delta_z c^j(x,z) + D_z \rho(z) D_z c^j(x,z)^T\} dz,$$

$$\bar{b}^{is}(x,y,u,v) = \frac{1}{2} \sum_{\ell,r=1}^{d} \frac{\partial^2 b^{js}}{\partial x_\ell \partial x_r}(x) u^{\ell r}$$

$$+ \sum_{\ell=1}^{d} \frac{\partial b^{js}}{\partial x_\ell}(x) v^\ell - \frac{1}{2} b^{js}(x),$$

$$\bar{c}^i((x,y,u,v),z) = \frac{1}{2} \sum_{\ell,r=1}^{d} \frac{\partial^2 c^j}{\partial x_\ell \partial x_r}(x,z) u^{\ell r}$$

$$+ \sum_{\ell=1}^{d} \frac{\partial c^j}{\partial x_\ell}(x,z) v^\ell + \frac{1}{2}\{\rho(z) \Delta_z c^j(x,z) + D_z \rho(z) D_z c^j(x,z)^T\}.$$

Now, we grade F exactly as in the proof of 6-42, and then a simple examination shows that these coefficients are graded according to this grading, and that they fulfill $(\tilde{A}'-(r-2))$. Since 10-20 and 10-21 do not depend on y, (X,U,V) satisfies also an equation 6-2 with $(\widetilde{A}'-(r-2))$.

Next, $(X,\nabla X,U)$ satisfies again an equation 6-2 whose coefficients are graded, and since there is no second derivative in 10-18, 10-19, 10-20, these coefficients fulfill $(\widetilde{A}'-(r-1))$. It is the same for (X,U), for a similar reason.

Proof of Theorem 10-3. We have just seen that $\hat{X}=(X,U,V)$ satisfies an equation 6-2 whose coefficients $(\hat{a},\hat{b},\hat{c})$ meet $(\widetilde{A}'-1)$. Let $\hat{X}^n=(X^n,U^n,V^n)$. A close comparison between 10-4 and 10-13, and between 10-5 and 10-14, shows that $\hat{X}^n$ satisfies an equation 5-27 with the following coefficients:

- If $i\leq d$ or $i>d+d^2$: $\hat{A}^{n,i}(y,\omega,t)=\hat{a}^i(y)$, $\hat{B}^{n,is}(y,\omega,t)=\hat{b}^{is}(y)$, $\hat{C}^{n,i}(y,\omega,t,z)=\hat{c}^i(y,z)$.

- If $d<i\leq d+d^2$ and i corresponds to the $(j,k)^{th}$ component in $\mathbb{R}^d\otimes\mathbb{R}^d$ and $y=(x,u,v)\in\mathbb{R}^d\times(\mathbb{R}^d\otimes\mathbb{R}^d)\times\mathbb{R}^d$:

$$\hat{A}^{n,i}(y,\omega,t) = \hat{a}^i(y) + \sum_{\ell,r=1}^{d}\{K_{t-}^{n,j\ell}(\omega)u^{\ell r}\frac{\partial a^k}{\partial x_r}(x) + \frac{\partial a^j}{\partial x_\ell}(x)u^{\ell r}K_{t-}^{n,kr}(\omega)\},$$

$$\hat{B}^{n,is}(y,\omega,t) = \hat{b}^{is}(y)+\sum_{\ell,r=1}^{d}\{K_{t-}^{n,j\ell}(\omega)u^{\ell r}\frac{\partial b^{ks}}{\partial x_r}(x) + \frac{\partial b^{js}}{\partial x_\ell}(x)u^{\ell r}K_{t-}^{n,kr}(\omega)\},$$

$$\hat{C}^{n,i}(y,\omega,t,z)=\hat{c}^i(y,z)+\sum_{\ell,r=1}^{d}\{K_{t-}^{n,j\ell}(\omega)u^{\ell r}\frac{\partial c^k}{\partial x_r}(x,z) + \frac{\partial c^j}{\partial x_\ell}(x,z)u^{\ell r}K_{t-}^{n,kr}(\omega)\}.$$

Since $(\hat{a},\hat{b},\hat{c})$ meets $(\widetilde{A}'-1)$, we easily deduce from what precedes that $(\hat{A}^n,\hat{B}^n,\hat{C}^n)$ satisfies 5-29, with

$$Z_t^n = \zeta+\zeta|K_{t-}^n|, \quad Z_t'^n = \zeta|K_{t-}^n|,$$

$$Z_t''^n = \sup_{r\leq q, M_{r-1}<i,j\leq M_r} \zeta|K_{t-}^{n,ij}|,$$

for some constant ζ. In view of Lemma 10-9, we see that $\sup_n |||Z^n|||_p<\infty$ and $|||Z'^n|||_p \to 0$ and $\beta_p<\infty$ if β_p is defined by 5-30. Hence 5-31 yields that

$$\|(\hat{X}^n-\hat{X})_T^*\|_{L^p} \to 0 \quad \text{as } n\uparrow\infty \text{ for all } p<\infty.$$

In particular, $X_t^{n,i} \to X_t^i$, $\Gamma(X_t^{n,i},X_t^{n,j}) \to U_t^{ij}$ and $LX_t^{n,i} \to V_t^i$ in all L^p. Since $X_t^{n,i}\in H_\infty$ by 10-12, we deduce from 8-17 that $X_t^i\in H_\infty$ and $\Gamma(X_t^i,X_t^j)=U_t^{ij}$ and $LX_t^i=V_t^i$.

§10-b. EXPLICIT COMPUTATION OF U

As seen in Section 8 (see 8-10), the matrix $\Gamma(X_t^i,X_t^j)$ plays a particularly important role. So we proceed to "explicitely" compute the solution to 10-4.

From now on, we make explicit the dependence of X upon the initial condition, writing X^{x_0} for the solution to 6-2. Then the solutions to 10-4 and 10-5 also depend on x_0, namely U^{x_0} and V^{x_0}. We also write ∇X^{x_0} for the solution to 10-16 (or 6-30), and we use the notation introduced in §6-d, namely K^x (see 6-31), T_n^x and $\nabla X^n(n)$ (see 6-35).

10-22 PROPOSITION: *Assume $(\tilde{A}'$-3). Then for $T_{n-1}^x\leq t<T_n^x$,*

$$U_t^x = \nabla X^x(n)_t \{ U_{T_{n-1}^x}^x + \int_{T_{n-1}^x}^t \nabla X^x(n)_{s-}^{-1} bb^T(X_{s-}^x)\nabla X^x(n)_{s-}^{-1,T} ds$$
$$+\iint_{ET_{n-1}^x}^t \nabla X^x(n)_{s-}^{-1}[(I+D_xc)^{-1}D_zcD_zc^T(I+D_xc)^{-1,T}](X_{s-}^x,z)$$
$$\nabla X^x(n)_{s-}^{-1,T}\rho(z)\mu(ds,dz)\}\nabla X^x(n)_t^T$$

Proof. In the proof we drop the superscript "x". Let us call R $(=R^x)$ the process defined in 7-5, with only one Poisson measure $\mu_\alpha=\mu$. By 7-1, R is well defined, with finite variation. Moreover, $\nabla X(n)_-^{-1}$ is well defined on $[\![0,T_n[\![$ and is left-continuous, so the following makes sense for $T_{n-1}\le t<T_n$:

$$\overline{U}_t = U_{T_{n-1}} + \int_{T_{n-1}}^{t} \nabla X(n)_{s-}^{-1}\, dR_s\, \nabla X(n)_{s-}^{-1,T}. \tag{10-23}$$

Furthermore, recalling 2-10 and 7-5, we observe that the right-hand side of the claimed formula is, for $T_{n-1}\le t<T_n$:

$$U(n)_t = \nabla X(n)_t\, \overline{U}_t\, \nabla X(n)_t^T. \tag{10-24}$$

Next, we set

$$H = (bb^T)(X_-)*t + \rho(D_z c.D_z c^T)(X_-)*\mu \tag{10-25}$$

(again this is well defined, because of $(\tilde{A}'$-3) and the property $\rho\in L^1(E,G)$). Then, comparing (10-25) and 7-5 and 10-23, and using the definition 6-31 of K $(=K^x)$, we easily obtain for $T_{n-1}\le t<T_n$:

$$\overline{U}_t = U_{T_{n-1}} + \int_{T_{n-1}}^{t} \nabla X(n)_{s-}^{-1}\,(I+\Delta K_s)^{-1}\, dH_s\,(I+\Delta K_s)^{-1,T}\,\nabla X(n)_{s-}^{-1,T}. \tag{10-26}$$

Moreover, we deduce from 10-4 and 6-31 and 10-25 that, still for $T_{n-1}\le t<T_n$:

$$\begin{aligned} U_t = {} & U_{T_{n-1}} + (H_t-H_{T_{n-1}}) + \int_{T_{n-1}}^{t} (U_{s-}dK_s^T + dK_s\, U_{s-}) \\ & + \sum_{i=1}^{m} \int_{T_{n-1}}^{t} D_x b^{\cdot i}(X_{s-})U_{s-}D_x b^{\cdot i}(X_{s-})ds \\ & + \sum_{T_{n-1}<s\le t} \Delta K_s U_{s-}\Delta K_s^T. \end{aligned} \tag{10-27}$$

Now we can apply Ito's formula to the triple product in 10-24, on the open interval $]\!]T_{n-1},T_n[\![$, thus getting for $T_{n-1} \le t < T_n$ (and with $[Z,Z']^c$ denoting the continuous part of the quadratic co-variation process between Z and Z'):

$$U(n)_t = U_{T_{n-1}} + \int_{T_{n-1}}^{t} \{\nabla X(n)_{s-}\overline{U}_{s-}d(\nabla X(n))_s^T$$

$$+ \nabla X(n)_{s-}d\overline{U}_{s-}\nabla X(n)_{s-}^T + d(\nabla X(n))_s\overline{U}_{s-}\nabla X(n)_{s-}^T\}$$

$$+ \sum_{i,j=1}^{d} \int_{T_{n-1}}^{t} \overline{U}_{s-}^{ij}\, d[\nabla X(n)^{\cdot i},\nabla X(n)^{\cdot j}]_s^c$$

$$+ \sum_{T_{n-1}<s\le t}\{[\nabla X(n)_s+\Delta(\nabla X(n))_s](\overline{U}_{s-}+\Delta\overline{U}_s)$$

$$[\nabla X(n)_{s-}+\Delta(\nabla X(n))_s]^T$$

$$- \nabla X(n)_{s-}\overline{U}_{s-}\nabla X(n)_{s-}^T - \nabla X(n)_{s-}\overline{U}_{s-}\Delta(\nabla X(n))_s^T$$

$$- \nabla X(n)_{s-}\Delta\overline{U}_s\nabla X(n)_{s-}^T - \Delta(\nabla X(n)_s\overline{U}_{s-}\nabla X(n)_{s-}^T\}.$$

Then 6-35, 6-31, 10-25 and 10-26 yield for $T_{n-1} \le t < T_n$:

$$U(n)_t = U_{T_{n-1}} + \int_{T_{n-1}}^{t} \{U(n)_{s-}dK_s^T + dK_sU(n)_{s-}\}$$

$$+ \int_{T_{n-1}}^{t} bb^T(X_{s-})ds + \sum_{T_{n-1}<s\le t}(I+\Delta K_s)^{-1}\Delta H_s(I+\Delta K_s)^{-1,T}$$

$$+ \sum_{i=1}^{m} \int_{T_{n-1}}^{t} D_xb^{\cdot i}(X_{s-})\; U(n)_{s-}\; D_xb^{\cdot i}(X_{s-})^T\, ds$$

$$+ \sum_{T_{n-1}<s\le t}\{\Delta K_s(I+\Delta K_s)^{-1}\Delta H_s(I+\Delta K_s)^{-1,T}\Delta K_s^T$$

$$+ (I+\Delta K_s)^{-1}\Delta H_s(I+\Delta K_s)^{-1,T}\Delta K_s^T$$

$$+\Delta K_s(I+\Delta K_s)^{-1}\Delta H_s(I+\Delta K_s)^{-1,T} + \Delta K_sU(n)_{s-}\Delta K_s^T\}$$

$$= U_{T_{n-1}} + \int_{T_{n-1}}^{t} \{U(n)_{s-} dK_s^T + dK_s U(n)_{s-}\} + (H_t - H_{T_{n-1}})$$
$$+ \sum_{i=1}^{m} \int_{T_{n-1}}^{t} D_x b^{i\cdot}(X_{s-}) U(n)_{s-} D_x b^{i\cdot}(X_{s-}) ds$$
$$+ \sum_{T_{n-1} < s \leq t} \Delta K_s \; U(n)_{s-} \; \Delta K_s^T$$

(we use the property

$$H_t - H_{T_{n-1}} = \sum_{T_{n-1} < s \leq t} \Delta H_s + \int_{T_{n-1}}^{t} bb^T(X_{s-}) ds).$$

This is a linear equation in $U(n)$, and it is idantical to Equation 10-27 in U : hence $U(n)_t = U_t$ for $T_{n-1} \leq t < T_n$.

§10-c. APPLICATION TO EXISTENCE AND SMOOTHNESS OF THE DENSITY

This subsection is similar to §6-f, in the sense that we prove here and there almost the same results, but presently with Malliavin-Stroock's approach instead of Bismut's one.

Firstly, we know that (L, H_∞) is a Malliavin operator (see 8-18) and we have described in 8-10 a related integration-by-parts setting for every random variable that belongs to H_∞. More precisely, under $(\tilde{A}'\text{-}3)$, the following $(\sigma_t^x, \gamma_t^x, \mathcal{H}_t^x, \delta_t^x)$ is an integration-by-parts setting for X_t^x:

$$\left.\begin{aligned} \sigma_t^x &= U_t^x, & \gamma_t^x &= -2V_t^x, \\ \mathcal{H}_t^x &= H_\infty, & \delta_t^{x,j}(\Psi) &= -\Gamma(X_t^{x,j}, \Psi) \text{ if } \Psi \in H_\infty. \end{aligned}\right\} \quad (10\text{-}28)$$

In order to apply the results of Section 4, we still have to describe the sets $C_{t,0}^x(q)$ introduced in it (see after 4-19). Consider the process

$$Y_0^x(q) = (U^x, V^x, \{\nabla^i X^x\}_{0\le i\le q\vee 1})$$

(so $Y_0^x(0)=Y_0^x(1)$), which by 6-42 is well defined for all $q\le r-1$, under $(\tilde{A}'-r)$ with $r\ge 3$. Then $C_{t,0}^x(q)$ is the set of all components of $Y_0^x(q)$ at time t, say $Y_{t,0}^x(q)$, and the iterates $C_{t,j}^x(q)$ are defined by 4-10. Recall that $Y_{t,j}^x(q)$ is a multi-dimensional variable whose components constitute $C_{t,j}^x(q)$.

10-29 LEMMA: *Assume* $(\tilde{A}'-r)$ *for some* $r\ge 5$. *Then*

a) $\{X_t^x\}$ *is* $(r-1)$ *times F-differentiable.*

b) $C_{t,r-4-q}^x(q)\subset H_\infty$ *for* $1\le q\le r-4$, *and* $C_{t,r-5}^x(0)\subset H_\infty$.

c) $x \rightsquigarrow \sup_{t\le T} E(|Y_{t,n}^x(q)|^p)$ *is locally bounded for all* $p<\infty$, *provided* $n+q\le r-3$ *if* $q\ge 1$, *and* $n\le r-4$ *if* $q=0$.

Proof. Set $Z^{0,x} = Y_0^x(1)$, as defined above. Lemma 10-17 yields that $Z^{0,x}$ satisfies an equation 6-2 with $(\tilde{A}'-(r-2))$. Since $r\ge 5$, the components of $Z_t^{0,x}$ are in H_∞ by 10-3, and are F-differentiable in x by 6-29, and we consider $Z^{1,x} = (Z^{0,x}, \nabla Z^{0,x}, \Gamma(Z^{0,x}, Z^{0,x}))$. Then by Lemma 10-17, $Z^{1,x}$ satisfies an equation 6-2 with $(\tilde{A}'-(r-3))$. If $r\ge 6$ we can pursue the construction, as such: if $Z^{k,x}$ is defined, and if it is a solution of an equation 6-2 with $(\tilde{A}'-(r-k-2))$, and if $k\le r-5$, then by 10-3 and 6-29 $Z^{k,x}$ is F-differentiable in x and its components are in H_∞: so we may set $Z^{k+1,x}= (Z^{k,x}, \nabla Z^{k,x}, \Gamma(Z^{k,x}, Z^{k,x}))$, which satisfies an equation 6-2 with $(\tilde{A}'-(r-k-3))$ from Lemma 10-17.

Now, (a) has already been proved, and (b) and (c) immediately follow from what precedes, once noticed that $Y_{t,0}^x(q)$ is just a subfamily of the components of $Z_t^{q-1,x}$, so an induction shows that $Y_{t,j}^x(q)$ is a subfamily of the components of $Z_t^{j+q-1,x}$ (for $q\geq 1$), or of $Z_t^{j,x}$ (for $q=0$) (for (c), the argument is the same as at the end of the proof of 6-47).

10-30 <u>THEOREM</u>: *Assume* $(\tilde{A}'-j)$ *and* 10-1, *and set* $Q_t^x = \det(U_t^x)$, *where* U^x *is the solution to* 10-4, *and*

$$q_t^x(i) = E(|Q_t^x|^{-i}) \quad (=\infty \quad \text{if } P(Q_t^x=0)>0) \tag{10-31}$$

a) If $j\geq 4$ *and* $Q_t^x\neq 0$ *a.s.,* X_t^x *admits a density* $y \rightsquigarrow p_t(x,y)$.

b) Moreover $p_t(x,.)$ *is of class* C^r, *provided:*

- *either* $j\geq r+d+5$ *and* $q_t^x(2r+2d+2+\varepsilon)<\infty$ *for some* $\varepsilon>0$,
- *or* $j\geq r+5$ *and* $q_t^x(2d(r+1)+\varepsilon)<\infty$ *for some* $\varepsilon>0$,

c) Moreover $(x,y)\rightsquigarrow p_t(x,y)$ *is of class* C^r, *provided:*

- *either* $j\geq r+2d+5$ *and* $x\rightsquigarrow q_t^x(2r+4d+2+\varepsilon)$ *is locally bounded for some* $\varepsilon>0$,
- *or* $j\geq r+5$ *and* $x \rightsquigarrow q_t^x(4d(r+1)+\varepsilon)$ *is locally bounded for some* $\varepsilon>0$.

d) Moreover,

(i) If $j\geq 2r+4d+8$, *if* $\sup_{t>t_0,x\in A} q_t^x(4r+8d+8+\varepsilon)<\infty$ *for every bounded subset* A *and some* $\varepsilon>0$ *(depending on* A*), and if*

$$|\det[I+vD_x c(x,z)]| \geq \zeta \qquad \forall v\in[0,1] \tag{10-32}$$

for some constant ζ, *then* $(t,x,y)\rightsquigarrow p_t(x,y)$ *is of class* C^r *on* $(t_0,T]\times\mathbb{R}^d\times\mathbb{R}^d$.

(ii) If $j\geq 2r+6$, *if* $\sup_{t>t_0,x\in A} q_t^x(4(r+1)(2d+1)+\varepsilon)<\infty$ *for every bounded set* A *and some* $\varepsilon>0$ *(depending on* A*),*

and if $c \equiv 0$, *then* $(t,x,y) \rightsquigarrow p_t(x,y)$ *is of class* C^r *on* $(t_0,T] \times \mathbb{R}^d \times \mathbb{R}^d$.

Proof. (a) follows from Theorem 4-7, once noticed that under $(\widetilde{A}'\text{-}4)$ the process (X^x, U^x) satisfies an equation 6-2 with $(\widetilde{A}'\text{-}3)$ by Lemma 10-17. (b), (c), (d) follow from Theorems 4-19, 4-21, 4-31, plus Lemma 10-29 and Lemma 6-51.

10-33 REMARK: Compare this with Theorem 6-48: of course Q_t^x is not the same variable in both theorems, but we shall see that the estimates on $q_t^x(i)$ when $Q^x = \det(U^x)$, in the next section, are the same as they were when $Q^x = \det(DX^x)$ in Section 7.

However, one needs one more (resp. two more) degrees of differentiability on the coefficients (a,b,c) in 10-30-a (resp. 10-30-b,c,d) than in 6-48-a (resp. 6-48-b,c,d). Furthermore, we need 10-1 (stronger than 6-9), and $(\widetilde{A}'\text{-}r)$ instead of $(A'\text{-}r)$. Hence Theorem 6-48 is (slightly) better than Theorem 10-30.

Section 11: PROOF OF THE MAIN THEOREMS VIA MALLIAVIN'S APPROACH

§11-a. INTRODUCTORY REMARKS

We want here to deduce Theorems 2-14, 2-27, 2-28, 2-29 from Theorem 10-30. And, exactly as in Section 7, we need to extend the setting of Sections 9 and 10 to encompass the situation of Section 2.

So, we consider the canonical setting of §7-a-1, supporting $W, (\mu_\alpha)_{\alpha \leq A}, \mu$. As in Section 10, the needed regularity conditions on the coefficients are slightly more than (A'-r), namely:

11-1 ASSUMPTION ($\tilde{A}$'-r): *The same as (A'-r) in 7-1, except that* $\eta_\alpha \in \cap_{2 \leq p \leq \infty} L^p(E_\alpha, G_\alpha)$ *for all* $\alpha = 1, \ldots, A$.

For each $\alpha \leq A$ we also consider a function ρ_α: $E_\alpha \to [0,\infty)$ satisfying 10-1 (and thus 9-17 as well).

Now, for translating Sections 8 and 9 the most convenient way consists in aggregating all measures μ_α and μ into a "big" measure $\overline{\mu} = \sum \mu_\alpha + \mu$, which is a Poisson measure on $[0,T] \times \overline{E}$, where $\overline{E} = \sum_\alpha E_\alpha + E$ ("disjoint" union). Then one considers Ω as being the canonical space accomodating W and $\overline{\mu}$. And the auxiliary function $\overline{\rho}$ which serves to constructing the Malliavin operator is

$$\overline{\rho} = \rho_\alpha \text{ on } E_\alpha, \quad \overline{\rho} = 0 \text{ on } E. \tag{11-2}$$

Obviously, all of Sections 9 and 10 carries over without modification, with W and $\overline{\mu}$. If we then come back to the original measures μ_α and μ, the fundamental formula 10-4 becomes

$$
\begin{aligned}
U^x = {} & bb^T(X^x_-)*t + \sum_\alpha \rho_\alpha (D_z c_\alpha)(D_z c_\alpha)^T (X^x_-)*\mu_\alpha \\
& + \{U^x_- D_x a(X^x_-)^T + D_x a(X_-)U^x_-\}*t \\
& + \sum_{i=1}^m \{U^x D_x b^{\cdot i}(X^x_-)^T + D_x b^{\cdot i}(X^x_-)U^x_-\}*W^i \\
& + \sum_{i=1}^m D_x b^{\cdot i}(X^x_-)U^x_- D_x b^{\cdot i}(X^x_-)^T*t \\
& + \sum \{U^x_- D_x c_\alpha (X^x_-)^T + D_x c_\alpha (X^x_-)U^x_-\}*\tilde{\mu}_\alpha \\
& + \{U^x_- D_x c(X^x_-)^T + D_x c(X^x_-)U^-_x\}*\tilde{\mu} + D_x c(X^x_-)U^x_- D_x c(X^x_-)^T*\tilde{\mu} \\
& + \sum_\alpha D_x c_\alpha (X^x_-)U^x_- D_x c_\alpha (X^x_-)^T*\tilde{\mu}_\alpha
\end{aligned}
$$

(note that $D_z c$, which does not exist, does not appear either, because $\bar{\rho}=0$ on E!) Finally, if we use the process R^x of 7-5, the explicit formula 10-22 (or rather 10-23 and 10-24) become in this context (again because $\bar{\rho}=0$ on E):

$$
U^x_t = \nabla X^x(n)_t \{U^x_{T^x_{n-1}} + \int_{T^x_{n-1}}^t \nabla X^x(n)^{-1}_{s-} dR^x_s \nabla X^x(n)^{-1,T}_{s-}\}.
$$

$$
.\nabla X^x(n)^T_t \quad \text{if } T^x_{n-1} \le t < T^x_n , \tag{11-3}
$$

(here T^x_n is given by 6-35, with K^x given by 7-2).

§11-b. EXISTENCE OF THE DENSITY

As observed in Remark 10-33, Theorem 10-30 is weaker than Theorem 6-48. So we will not prove 2-14 in full generality. Rather, we will assume ($\widetilde{A}$'-4) instead of (A'-3) (and, of course, we assume (B)).

Since E_α is a countable union of β_α-dimensional rectangles, it is easy to find a function ρ_α which meets both 10-1 and $\rho_\alpha>0$ on E_α.

In virtue of 10-30-a, it suffices to prove that for

every $t\in(0,T]$, U_t^x is a.s. invertible. Since T_n^x is the time of jump of one of the Poisson measures μ_α or μ, we have $P(T_n^x=t)=0$, so it suffices to prove that U_t^x is a.s. invertible on every set $A_n^x = \{T_{n-1}^x<t<T_n^x\}$.

On this set A_n^x, U_t^x is given by 11-3, while $\nabla X^x(n)_t$ is invertible, so it is enough to prove that

$$U^x_{T^x_{n-1}} + \int_{T^x_{n-1}}^t \nabla X^x(n)_{s-}^{-1}\, dR_s^x\, \nabla X(n)_{s-}^{-1,T}$$

is a.s. invertible. This being the sum of two symmetric nonnegative matrices, it suffices that the second one be a.s. invertible. But this immediately follows from Lemma 7-9, and we are finished.

§11-c. SMOOTHNESS OF THE DENSITY

Here we assume (SB-(ζ,θ)) and (SC) and at least ($\tilde{A}$'-5). In addition we suppose that each function ρ_α in (SB-(ζ,θ)) meets 9-17, and we use these functions to construct the Malliavin operator. Because of (SC), $T_1^x=\infty$ a.s., therefore 11-3 and 7-5 yield

$$U_t^x = \nabla X_t^x\, S_t^x\, \nabla X_t^{x,T}, \tag{11-4}$$

where S^x is given by 7-32, with $k=1$, $k'_\alpha=1$. Since these functions meet 7-33, we deduce that 7-34 and thus 7-50 hold. Applying 11-4 and Lemma 7-27 allows to deduce that 7-51 holds, where $q_s^x(i)=E(\det(U_s^x)^{-i})$ (exactly as in 7-47).

Therefore we deduce Theorems 2-27, 2-28 and 2-29 from 10-30-b,c,d respectively, with the following restrictions: we need ($\tilde{A}$'-(i+2)) instead of (A'-i), and ρ_α should meet 9-17.

Section 12: CONCLUDING REMARKS

1 - About the Malliavin operator on Poisson space. If the conditions 9-17 on ρ are not fulfilled, in order that the conclusion of Theorem 9-18 be true it is necessary to strengthen the hypotheses 9-16 of f and F. Then in turn, the solution to Equation 6-2 do not belong to the domain H_∞, unless we considerably strengthen $(\tilde{A}'-2)$: for instance, $\sup_x |D_z c(x,.)|$ should not only be bounded, but also in $L^2(E,G)$.

As a matter of fact, one could replace the function ρ by

$$\tilde{\rho} \text{ is a } C^1 \text{ function on E, with values in the set of } \beta\times\beta \text{ symmetric nonnegative matrices.} \tag{12-1}$$

Then in 9-3 we should change the following:

$$\left.\begin{aligned} &\tfrac{1}{2}(\rho\Delta_z f_i + D_z\rho(D_z f_i)^T) \text{ replaced by} \\ &\quad \frac{1}{2}\sum_{k,\ell=1}^{\beta}\left(\tilde{\rho}^{k\ell}\frac{\partial^2}{\partial z_k \partial z_\ell} f_i + \frac{\partial\tilde{\rho}^{k\ell}}{\partial z_k}\frac{\partial f_i}{\partial z_\ell}\right) \\ &\rho D_z f_i (D_z f_j)^T \text{ replaced by } \sum_{k,\ell=1}^{\beta}\tilde{\rho}^{k\ell}\frac{\partial f_i}{\partial z_k}\frac{\partial f_j}{\partial z_\ell} \end{aligned}\right\} \tag{12-2}$$

and everything else would work fine. We could even make $\tilde{\rho}$ depend on t. However, these generalizations do not seem to improve on the results obtained for stochastic differential equations.

2 - The Malliavin operator is an infinitesimal generator. Likewise the proper Malliavin operator on Wiener

space, which is the generator of an "infinite dimensional Ornstein-Uhlenbeck process" (see [19], [24], [25]) with values in $C([0,T];\mathbb{R}^m)$, we can interpret the operator L constructed in §9-a as the generator of a Markov process $M=(M_s)_{s\geq 0}$ as follows (rather, (L,H_∞) is the restriction to H_∞ of the generator).

The state space of M is the canonical space $(\Omega,\underline{\underline{F}})$ of §9-a. Knowing the initial value $M_0=\mu$, the dynamics of M is as follows: μ is a point measure whose support can be written as $\{(t,z(t):t\in D\}$ where D is countable; for each $t\in D$ one runs a diffusion process $(Z_s(t))_{s\geq 0}$ on E, with generator

$$\frac{1}{2}\{\rho(z)\Delta_z f + D_z\rho(z).D_z f^T\}$$

and starting at $Z_0(t)=z(t)$, and all these diffusion processes are independent. Then set

M_s is the point measure with support
$\{(t,Z_s(t)):t\in D\}$.

Then $M=(M_s)_{s\geq 0}$ is clearly Markov. Each process $Z(t)$ is reversible with respect to Lebesgue measure on E, which implies that M admits the canonical Poisson measure P as a stationary measure and that it is reversible under this initial measure: this corresponds to the self-adjoint property for L. Also abserve that 9-17 implies that each diffusion process $Z_\cdot(t)$ lives inside E and never reaches the boundary.

If we replace ρ by $\tilde{\rho}$, according to 12-1, M is constructed similarly, but the generator of each $Z(t)$ should be modified according to the first formula in 12-2.

This sort of point measure-valued Markov processes

is of course well known in other contexts: see for example Surgailis [27].

3 - A differential operator on the Poisson space. When $(\Omega,\underline{\underline{F}},P)$ is the Wiener space of §9-b, let H be the Hilbert space of absolutely continuous $\mathbb{R}^m$-valued functions on $[0,T]$ with Lebesgue square-integrable derivatives, endowed with the usual scalar product. Then Shigekawa [23] introduces the derivative $\mathcal{D}\Phi$ of $\Phi\in\mathcal{R}$ as being the "Fréchet derivative along H"; then he defines Γ as

$$\Gamma(\Phi,\Psi) = \langle\mathcal{D}\Phi,\mathcal{D}\Psi\rangle_H$$

and then he defines L through Γ (it is closely related, but slightly more complicated than the approach of §9-b).

Let us come back to the Poisson space $(\Omega,\underline{\underline{F}},P)$ of §9-a. We also have a notion of derivative $\mathcal{D}$. More precisely let $\Phi=F(\mu(f_1),\ldots,\mu(f_k))$ be in $\mathcal{D}$. Then set

$$\mathcal{D}\Phi(\omega,s,z) = \sum_{i=1}^{k} \frac{\partial F}{\partial x_i}(\mu(f_1),\ldots,\mu(f_k))D_z f_i(s,z). \quad (12\text{-}3)$$

One may prove (as in 9-4) that $\Phi(\omega,.)$ is defined up to a $\mu(\omega;.)$-null set. Note that $\mathcal{D}\Phi$ is an $\mathbb{R}^\beta$-valued function on $\Omega\times[0,T]\times E$. Then one can show that $\Gamma(\Phi,\Psi)$ is

$$\Gamma(\Phi,\Psi) = \int_0^T\int_E \mathcal{D}\Phi(.,s,z)\mathcal{D}\Psi(.,s,z)^T\rho(z)\mu(ds,dz) \quad (12\text{-}4)$$

which, with obvious notation, can also be written as

$$\Gamma(\Phi,\Psi) = \langle\mathcal{D}\Phi,\mathcal{D}\Psi\rangle_{\rho\mu} \quad (12\text{-}5)$$

($\mathcal{D}$ is not a Fréchet-type derivative, since Ω is not a linear space; however, Ω can be viewed as an infinite-dimensional manifold and then one may interpret $\mathcal{D}$ as a derivative along subspaces of the tangent space).

4 - Comparison of the two approaches. We have already emphazised the differences a number of times, and also discussed the advantages of the first one (at least as long as smoothness problems for stochastic differential equations are concerned). The above-mentioned "derivative" $\mathcal{D}$ allows for a more thorough comparison. In the second approach the key role is played by

$$U_t^x = \langle \mathcal{D}X_t^x, \mathcal{D}X_t^x \rangle_{\rho\mu} \tag{12-6}$$

(suppose for simplicity that there is no Wiener process and that everything is 1-dimensional). In the first approach we use rather

$$DX_t^x = \langle \mathcal{D}X_t^x, v^x \rangle_{\rho\mu} \tag{12-7}$$

where v^x is the function on $\Omega\times[0,T]\times E$ introduced in 6-7 or 6-38. Note that in 12-7, v^x does not depend on t, but is predictable on $\Omega\times[0,T]\times E$, which is not the case of $\mathcal{D}X_t^x(\omega,s,z)$.

So it seems that, mutatis mutandis, the second approach automatically yields the "best" perturbation insuring that U_t^x is invertible, while in Bismut's approach we have to choose the best v^x upon examination of the explicit formula giving DX_t^x (Observe also that the proof of inversibility for U_t^x is significantly easier than for DX_t^x, in the course of proving Theorem 2-14).

REFERENCES

1. R.F. BASS, M. CRANSTON: The Malliavin calculus for pure jump processes, and applications to local time. Ann. Probab. 14, 490-532,(1986).
2. K. BICHTELER: Stochastic integrators with independent increments. Zeit. für Wahr. 58, 529-548,(1981).
3. K. BICHTELER, D. FONKEN: A simple version of the Malliavin calculus in dimension one. In Proc. Cleveland Conf. Mart. Theory. Lecture Notes in Math. 939 6-12,(1982), Springer Verlag: Berlin, Heidelberg, New-York.
4. K. BICHTELER, D. FONKEN: A simple version of the Malliavin calculus in dimension N. Seminar on Stoch. Processes (Evanston) 97-110,(1983). Birkhäuser: Boston.
5. K. BICHTELER, J. JACOD: Calcul de Malliavin pour les diffusions avec sauts, existence d'une densité dans le cas uni-dimensionnel. Séminaire de Proba. XVII. Lecture Notes in Math. 986, 132-157,(1983), Springer Verlag: Berlin, Heidelberg, New-York.
6. J.M. BISMUT: Martingales, the Malliavin calculus, and hypoellipticity under general Hörmander conditions. Zeit. für Wahr. 56, 469-505,(1981).
7. J.M. BISMUT: Calcul des variations stochastiques et processus de sauts. Zeit. für Wahr. 63, 147-235, (1983).
8. J.M. BISMUT: The calculus of boundary processes. Ann. Ecole Norm. Sup. 17, 507-622,(1984).
9. D. FONKEN: A simple version of Malliavin calculus with applications to the filtering theory, (1984).
10. J.B. GRAVEREAUX, J. JACOD: Opérateur de Malliavin sur l'espace de Wiener-Poisson. Compte R. Acad. Sci. 300, 81-84,(1985).
11. U. HAUSSMANN: On the integral representation of Ito processes. Stochastics 3, 17-27,(1979).
12. N. IKEDA, S. WATANABE: Stochastic differential equations and diffusion processes. North Holland (1979), Amsterdam.

13. J. JACOD: Calcul stochastique et problèmes de martingales, Lecture Notes in Math. 714 (1979), Springer Verlag: Berlin, Heidelberg, New-York.

14. J. JACOD: Equations différentialles linéaires, la méthode de variation des constantes. Séminaire Proba. XVI, Lecture Notes in Math. 920, 442-458,(1982), Springer Verlag: Berlin, Heidelberg, New-York.

15. H. KUO: Brownian functionals and applications. Acta Appl. Math. 1, 1-14,(1983).

16. S. KUSUOKA, D. STROOCK: Applications of the Malliavin calculus, Part I. Proc. 1982 Int'l Conf. Katata, Kinokuniya Publ. Co.: Tokyo.

17. R. LEANDRE: Régularité des processus de sauts dégénérés, Ann. Inst. H. Poincaré 21, 125-146,(1985).

18. R. LEANDRE: Thèse 3ème cycle, Besançon (1984).

19. P. MALLIAVIN: Stochastic calculus of variations and hypoelliptic operators. Proc. Int'l Conf. on Stoch. Diff. Equa., Kyoto 1976, 195-263. Wiley (1978): New-York.

20. P.A. MEYER: Un cours sur les intégrales stochastiques, Séminaire Proba X, Lecture Notes in Math 511, Springer Verlag: Berlin, Heidelberg, New-York.

21. J. NORRIS: Simplified Malliavin calculus. To appear in: Séminaire Proba. XX.

22. H. RUBIN: Supports of convolutions of identical distributions, Proc. 5th Berkeley Symp.II/1, 415-422, (1967). Univ. Calif. Press: Berkeley.

23. I. SHIGEKAWA: Derivatives of Wiener functionals and absolute continuity of induced measures, J. Kyoto Univ. 20, 263-289,(1980).

24. D. STROOCK: The Malliavin calculus and its applications to second order parabolic differential equations, Math. Systems Theory 14, 25-65 and 141-171, (1981).

25. D. STROOCK: The Malliavin calculus and its applications. In Stochastic Integrals (D. Williams ed.), Lecture Notes in Math. 851, 394-432, (1981), Springer Verlag: Berlin, Heidelberg, New-York.

26. D. STROOCK: The Malliavin calculus, a functional analytic approach. J. Funct. Analysis 44, 212-258, (1981).

27. D. SURGAILIS: On Poisson multiple integrals and associated equilibrium Markov processes. In Theory and Applications of random fields, Lecture Notes in Control and Inf. Sci. 49, 233-248, (1983), Springer Verlag: Berlin, Heidelberg, New-York.

28. H.G. TUCKER: Absolute continuity od infinitely divisible distributions, Pacific J. Math. 12, 1125-1129, (1962).

29. M. ZAKAI: The Malliavin calculus, Acta Appl. Math. (to appear).

INDEX

NOTATION